U.S.NRC
United States Nuclear Regulatory Commission

Protecting People and the Environment

NUREG-1855, Rev. 1

Guidance on the Treatment of Uncertainties Associated with PRAs in Risk-Informed Decisionmaking

Draft Report for Comment

Office of Nuclear Regulatory Research

AVAILABILITY OF REFERENCE MATERIALS
IN NRC PUBLICATIONS

NUREG-1855, Rev. 1

Guidance on the Treatment of Uncertainties Associated with PRAs in Risk-Informed Decisionmaking

Draft Report for Comment

Manuscript Completed: January 2013
Date Published: March 2013

Prepared by:
M. Drouin[1], A. Gilbertson[1], G. Parry[2]
J. Lehner[3], G. Martinez-Guridi[3]
J. LaChance[4], T. Wheeler[4]

[1]Office of Nuclear Regulatory Research
[2]Former NRC Employee
[3]Brookhaven National Laboratory
 Upton, NY 11973
[3]Sandia National Laboratories
 Albuquerque, NM 87185

M. Drouin, NRC Project Manager

Office of Nuclear Regulatory Research

COMMENTS ON DRAFT REPORT

Any interested party may submit comments on this report for consideration by the NRC staff. Comments may be accompanied by additional relevant information or supporting data. Please specify the report number "NUREG-1855 Revision 1, Draft" in your comments, and send them to the following address by the end of the comment period.

Chief, Rules, Announcements, and Directives Branch
Division of Administrative Services
Office of Administration
Mail Stop TWB-05-B01
U.S. Nuclear Regulatory Commission
Washington, DC 20555-0001

Comments may be submitted electronically using the NRC's website:
http://www.nrc.gov/public-involve/doc-comment/form.html

For any questions about the material in this report, please contact:

Mary Drouin
Mail Stop: CSB-4A07m
U.S. Nuclear Regulatory Commission
Washington, DC 20555-0001
Phone: (301) 251-7574
E-mail: Mary.Drouin@nrc.gov

Please be aware that any comments that you submit to the NRC will be considered a public record and entered into the Agencywide Documents Access and Management System (ADAMS). Do not provide information you would not want to be publicly available.

ABSTRACT

This document provides guidance on how to treat uncertainties associated with probabilistic risk assessment (PRA) in risk-informed decisionmaking. The objectives of this guidance include fostering an understanding of the uncertainties associated with PRA and their impact on the results of PRA and providing a pragmatic approach to addressing these uncertainties in the context of the decisionmaking.

In implementing risk-informed decisionmaking, the U.S. Nuclear Regulatory Commission (NRC) expects that appropriate consideration of uncertainty will be given in the analyses used to support the decision and in the interpretation of the findings of those analyses. To meet the objective of this document, it is necessary to understand the role that PRA results play in the context of the decision process. To define this context, this document provides an overview of the risk-informed decisionmaking process itself.

With the context defined, this document describes the characteristics of a risk model and, in particular, a PRA. This description includes recognition that a PRA, being a probabilistic model, characterizes aleatory uncertainty that results from randomness associated with the events of the model. Because the focus of this document is epistemic uncertainty (i.e., uncertainties in the formulation of the PRA model), it provides guidance on identifying and describing the different types of sources of epistemic uncertainty and the different ways that they are treated. The different types of epistemic uncertainty are parameter, model, and completeness uncertainties.

The final part of the guidance addresses the uncertainty in PRA results in the context of risk-informed decisionmaking and, in particular, the interpretation of the results of the uncertainty analysis when comparing PRA results with the acceptance guidelines established for a specified application. In addition, guidance is provided for addressing completeness uncertainty in risk-informed decisionmaking. Such consideration includes using a program of monitoring, feedback, and corrective action.

FOREWORD

In its safety philosophy, the U.S. Nuclear Regulatory Commission (NRC) always has recognized the importance of addressing uncertainties as an integral part of its decisionmaking. With the increased use of probabilistic risk assessment (PRA) in its risk-informed decisionmaking, the NRC needs to consider the uncertainties associated with PRA. Moreover, to ensure that informed decisions are made, the NRC must understand the potential impact of these uncertainties on the comparison of PRA results with acceptance guidelines. When dealing with completeness uncertainties, the NRC also must understand the use of bounding analyses to address potential risk contributors not included in the PRA. Ultimately, when addressing uncertainties in risk-informed decisionmaking, the NRC must develop guidance for those uncertainties associated with the PRA and those associated with our state of knowledge regarding design, etc., and the different decisionmaking processes (e.g., expert panel). In addition, this guidance should cover the various stages of the plant that the PRA is assessing (i.e., design, construction, and operation); the different types of reactors (e.g., light water reactors (LWRs) and non-LWRs); the different risk metrics (e.g., core damage frequency, radionuclide release frequency); and the different plant-operating states and hazard groups.

At this time, the focus of the guidance developed is for addressing uncertainties associated with PRAs. Although the process discussed in this document is more generally applicable, the detailed guidance on sources of uncertainty is focused on those sources associated with PRAs assessing core-damage frequency and large early-release frequency for operating LWRs for all plant operational states and considering internal hazards and external hazards.

In initiating this effort, the NRC recognized that the Electric Power Research Institute (EPRI) also was performing work in this area with similar objectives. Both the NRC and EPRI believed a collaborative effort to have technical agreement and to minimize duplication of effort would be more effective and efficient. The NRC and EPRI have worked together under a Memorandum of Understanding and, as such, the two efforts complement each other.

As noted above, it is repeatedly stated that "the NRC needs to," or "the NRC must," however, it is equally important that licensees understand and appropriately address uncertainties in their risk-informed regulatory activities. This NUREG was revised to better structure guidance to the licensee versus describing the review process followed by the NRC staff in developing their risk-informed decision on a risk-informed application. Moreover, additional explanation was added to further clarify the guidance and staff review process.

TABLE OF CONTENTS

LIST OF FIGURES

LIST OF TABLES

ACKNOWLEDGMENTS

The U. S. Nuclear Regulatory Commission (NRC) and the Electric Power Research Institute (EPRI) have collaborated in their mutual efforts on this topic. As such, both this NUREG and the EPRI documents complement each other. The NRC would like to acknowledge, that as part of this collaboration, EPRI (i.e., Ken Canavan, Stuart Lewis, and Mary Presley) and its contractor ERIN Engineering (i.e., Don Vanover, Gareth Parry and Doug True) provided significant comments.

ACRONYMS AND ABBREVIATIONS

ac	alternating current
ACRS	Advisory Committee on Reactor Safeguards
ALWR	advanced light-water reactor
ANS	American Nuclear Society
CC	capability category
CDF	core damage frequency
CFR	*Code of Federal Regulations*
dc	direct current
EPRI	Electric Power Research Institute
FV	Fussell-Vesely
HEP	human error probability
HPCI	high-pressure coolant injection
HRA	human reliability analysis
ICCDP	incremental conditional core damage probability
ICDF	incremental core damage frequency
ICLERP	incremental conditional large early release probability
ILERF	incremental large early release frequency
ILERP	incremental large early release probability
LERF	large early release frequency
LHS	Latin hypercube sampling
LOCA	loss-of-coolant accident
LOOP	loss of offsite power
LPSD	low power and shutdown
LWR	light-water reactor
MCS	minimal cut set
MOV	motor-operated valve
NEI	Nuclear Energy Institute
NPP	nuclear power plant
NRC	U.S. Nuclear Regulatory Commission
pdf	probability distribution function
POS	plant operating state
PRA	probabilistic risk assessment
RAW	risk achievement worth
RCP	reactor coolant pump
RG	regulatory guide
SEM	standard error of the mean
SOKC	state-of-knowledge correlation
SR	supporting requirement
SSC	structure, system, and component
yr	year

1. INTRODUCTION

1.1 Background and History

In a 1995 policy statement [NRC, 1995a], the U.S. Nuclear Regulatory Commission (NRC) encouraged the use of probabilistic risk assessment (PRA) in all regulatory matters. The policy statement declares the following:

> the use of PRA technology should be increased to the extent supported by the state-of-the-art in PRA methods and data and in a manner that complements NRC's deterministic approach . . . and supports the NRC's traditional defense-in-depth philosophy . . . PRA and associated analyses (e.g., sensitivity studies, uncertainty analyses and importance measures) should be used in regulatory matters . . . where practical within the bounds of state-of-the-art, to reduce unnecessary conservatism associated with current regulatory requirements, regulatory guides, license commitments, and staff practices.

The Commission further notes the following in the 1995 policy statement

> treatment of uncertainty is an important issue for regulatory decisions. Uncertainties exist . . . from knowledge limitations . . . A probabilistic approach has exposed some of these limitations and provided a framework to assess their significance and assist in developing a strategy to accommodate them in the regulatory process.

In a white paper titled, "Risk-Informed and Performance-Based Regulation" [NRC, 1999], the Commission defined the terms and described its expectations for risk-informed and performance-based regulation. The Commission indicated that a "risk-informed" approach explicitly identifies and quantifies sources of uncertainty in the analysis (although such analyses do not necessarily reflect all important sources of uncertainty) and leads to better decisionmaking by providing a means to test the sensitivity of the results to key assumptions.

Since the issuance of the PRA policy statement, NRC has implemented or undertaken numerous uses of PRA including modification of its reactor safety inspection program and initiation of work to modify reactor safety regulations. Consequently, confidence in the information derived from a PRA is an important issue. The technical adequacy of the content has to be sufficient to justify the specific results and insights to be used to support the decision under consideration. The treatment of the uncertainties associated with the PRA is an important factor in establishing this technical acceptability. Deterministic analyses that are performed in licensing applications contain uncertainties; however, they are addressed via defense-in-depth and safety margin. As such, a systematic process that is used to identify and understand deterministic uncertainties is not always needed.

Regulatory Guide (RG) 1.200 [NRC, 2007a] and the PRA consensus standard published by ASME and the American Nuclear Society (ANS) [ASME/ANS, 2009] each recognize the importance of identifying and understanding uncertainties as part of the process of achieving technical acceptability in a PRA, and these references provide guidance on this subject. However, they do not provide explicit guidance on the treatment of uncertainties in risk-informed decisionmaking.

RG 1.200 states that a full understanding of the uncertainties and their impact is needed (i.e., sources of uncertainty are to be identified and analyzed). Specifically, RG 1.200 notes the following:

> An important aspect in understanding the base PRA results is knowing the sources of uncertainty and assumptions and understanding their potential impact. Uncertainties can be either parameter or model uncertainties, and assumptions can be related either to PRA scope and level of detail or to model uncertainties. The impact of parameter uncertainties is gained through the actual quantification process. The assumptions related to PRA scope and level of detail are inherent in the structure of the PRA model. The requirements of the applications will determine whether they are acceptable. The impact of model uncertainties and related assumptions can be evaluated qualitatively or quantitatively. The sources of model uncertainty and related assumptions are characterized in terms of how they affect the base PRA[1] model (e.g., introduction of a new basic event, changes to basic event probabilities, change in success criterion, introduction of a new initiating event).

For the various risk-informed activities, the sources of uncertainties need to be addressed. The actual treatment, however, can vary based on the approach used. For example, RG 1.174 states that a PRA should include a full understanding of the impacts of the uncertainties through either a formal quantitative analysis or a simple bounding or sensitivity analyses. RG 1.174 also maintains that the decisions "must be based on a full understanding of the contributors to the PRA results and the impacts of the uncertainties, both those that are explicitly accounted for in the results and those that are not."

The ASME/ANS standard on PRA[2] requires that both parameter and model uncertainties be addressed. For example, parameter uncertainties are addressed via the quantification process of the core damage and large early release frequencies and model uncertainties also have to be identified and characterized. However, regardless of whether the uncertainty is a parameter or model uncertainty, the standard only provides requirements that describe what to do to address those uncertainties, but not how to address them.

In a letter dated April 21, 2003 [ACRS, 2003a], the Advisory Committee on Reactor Safeguards (ACRS) provided recommendations for staff consideration in Draft Guide 1122 (now RG 1.200). One recommendation was to include guidance on how to perform sensitivity and uncertainty analyses. In response to the ACRS [NRC, 2003b], the Commission agreed that guidance is needed for the treatment of uncertainties in risk-informed decisionmaking (i.e., the role of sensitivities and uncertainty analyses). Specifically, guidance is needed regarding the acceptable characterization of other methods, such as bounding analyses, to ensure that

---

[1] A base PRA is the PRA model that estimates the risk of the as-built and as-operated plant independent of an application. It is the base PRA model that is revised, for example, to estimate the change in risk from a proposed design change.

[2] As of January 2011, the current version of the ASME/ANS PRA standard is ASME/ANS RA-Sa-2009 [ASME/ANS, 2009]. This standard is being maintained and future editions are anticipated. However, it is not expected that the requirements with regard to uncertainties will be revised; that is, it is expected that the ASME/ANS PRA standard will always require uncertainties to be identified and characterized.

credible approaches[3] are used. That guidance was provided in the original version of this report and in complimentary reports from the Electric Power Research Institute (EPRI).

NUREG-1855 was first issued for use in March 2009. Following publication, a major public workshop was held and a test case using the guidance in the NUREG was performed to assess the effectiveness of this guidance. Further, progress in implementation of the risk-informed activities has occurred. Insights from the public, the test case, and risk-informed activities identified numerous areas for improvement to the guidance and scope of this NUREG.

1.2   Objectives

This document provides guidance on how to treat uncertainties associated with PRAs used by a licensee or applicant to support a risk-informed application to NRC. Specifically, guidance is provided with regard to:

- identifying and characterizing the uncertainties associated with PRA

- performing uncertainty analyses to understand the impact of the uncertainties on the results of the PRA

- factoring the results of the uncertainty analyses into the decisionmaking

With regard to the first two objectives, ASME and the ANS have been developing standards on PRA that support these objectives. Specifically, the ASME/ANS PRA standard [ASME/ANS, 2009] provides requirements[4] related to identifying, characterizing, and understanding the impact of the uncertainties. However, the standard only specifies what needs to be done to address uncertainties. Before the publication of NUREG-1855, formal guidance had not yet been developed regarding how to meet these requirements or on how to include the above aspects of the treatment of uncertainties into the decisionmaking.

Furthermore, the guidance in this document is intended for both the licensee and the NRC. That is, guidance is provided with regard to (1) NRC expectations of how the licensee should address PRA uncertainties in the context of an application and (2) how the impact of those uncertainties is evaluated by the NRC in a risk-informed application.

Beyond the PRA standard requirements related to identifying, characterizing, and understanding the impacts of PRA uncertainties, the ASME/ANS PRA standard also requires a peer review of the PRA. Similarly, the standard only states what should be included in the peer review and not how it should be performed. PRA peer review guidance is provided in various documents provided by the Nuclear Energy Institute (NEI) and is endorsed in RG 1.200. This NUREG is also intended to support the PRA peer review guidance provided by standards development organizations and nuclear industry organizations [NEI, 2005a; NEI, 2006a; NEI, 2006b] as it relates to the treatment of uncertainties.

---

[3] Credibility is obtained when there is a sound technical basis such that the basis would receive broad acceptance within the relevant technical community. The relevant technical community includes those individuals with explicit knowledge of and experience with the given issue.

[4] The use of the word "requirement" is standards language (e.g., in a standard, it states that the standard "sets forth requirements") and the use of the word is not meant to imply a regulatory requirement.

Further, EPRI, in parallel with NRC, has developed guidance documents on the treatment of uncertainties. This NUREG and the EPRI guidance have been developed to complement each other and are intended to be used as such when assessing the treatment of uncertainties in PRAs used in risk-informed decisionmaking. Where applicable, the NRC guidance refers to the EPRI work for acceptable approaches for the treatment of uncertainties [EPRI, 2004; EPRI, 2006; EPRI, 2008; EPRI, 2012] (See Section 1.5).

1.3 Scope and Limitations

The guidance in this document focuses on acceptable methods for addressing uncertainties associated with the use of PRA insights and results used in risk-informed decisionmaking. This document does not provide guidance for uncertainties associated with other types of analyses that support a risk-informed application (e.g., deterministic analysis).

Although the guidance in the this report does not currently address all sources of uncertainty, the guidance provided on the uncertainty identification and characterization process and on the process of factoring the results into the decisionmaking is generic and independent of the specific source of uncertainty. Consequently, the guidance is applicable for sources of uncertainty in PRAs that address at-power and low power and shutdown operating conditions, and both internal and external hazards.

In addressing uncertainties, expert judgment or elicitation may be used to determine if an uncertainty exists as well as the nature of that uncertainty (e.g., magnitude). This NUREG does not provide guidance on the use of expert judgment or performing expert elicitation. Guidance on this subject can be found in NUREG/CR-6372 [LLNL, 1997] or NUREG-1563 [NRC, 1996].[5]

An expert panel may be convened to address significant risk contributors that are not covered by a standard. The use of an expert panel implicitly takes into consideration the sources of uncertainty associated with those risk contributors. This NUREG does not provide guidance on employing an expert panel.

This guidance has been developed with a focus on the fleet of currently operating reactors. In particular, some of the numerical screening criteria referred to in this document and the identification of the sources of model uncertainty included in the EPRI report documents [EPRI, 2004a; EPRI, 2006a] (see Section 1.5) are informed by experience with PRAs for currently operating reactors. Nonetheless, the process is applicable for advanced light-water reactors (ALWRs) and non-LWRs and reactors in the design stage; however, the screening criteria and the specific sources of uncertainty may not be applicable. Consequently, some sources of uncertainty unique to ALWRs and non-LWRs and reactors in the design stage will exist that are not addressed in this report or the EPRI reports.

In developing the sources of model uncertainty, a model uncertainty needs to be distinguished from an assumption or approximation that is made, for example, on the level of detail needed for a given risk-informed activity. Although assumptions and approximations can influence the decisionmaking process, they are generally not considered to be model uncertainties because the level of detail in the PRA model could be enhanced, if necessary. Therefore, methods for addressing this aspect are not explicitly included in this report; they are, however, addressed in the EPRI reports.

---

[5] In addition, NUREG/CR-4550, Vol. 2 provides the approach used in NUREG-1150 on expert elicitation.

1.4   Approach Overview

In developing the necessary guidance to meet the objectives on how to treat uncertainties associated with PRA in risk-informed decisionmaking, the guidance needs to achieve the following:

- identify the different types of uncertainties that need to be addressed
- address the treatment to be performed by the licensee/applicant
- address how the staff accounts for the treatment in their decisionmaking

Guidance is provided on addressing the different types of uncertainties and focuses on the type of uncertainty that need to be accounted for in the decisionmaking. Generally speaking, there are two main types of uncertainty; aleatory and epistemic. Aleatory uncertainty is based on the randomness of the nature of the events or phenomena and cannot be reduced by increasing the analyst's knowledge of the systems being modeled. Therefore, it is also known as random uncertainty or stochastic uncertainty. Epistemic uncertainty is the uncertainty related to the lack of knowledge about or confidence in the system or model and is also known as state-of-knowledge uncertainty.

PRA models explicitly address aleatory uncertainty which results from the randomness associated with the events of the model in the logic structure, and methods have been developed to characterize one type of epistemic uncertainty, namely parameter uncertainty. The focus of this document is epistemic uncertainty (i.e., uncertainties related to the lack of knowledge). This guidance provides acceptable methods of identifying and characterizing the different types of epistemic uncertainty and the ways that those uncertainties are treated. The different types of epistemic uncertainty are completeness, parameter, and model uncertainty.

- <u>Completeness Uncertainty</u> – Guidance is provided on how to address one aspect of the treatment of completeness uncertainty (i.e., missing scope) in risk-informed applications. This guidance describes how to perform a conservative or bounding analysis to address items missing from a plant's PRA scope.

- <u>Parameter Uncertainty</u> – Guidance is provided on how to address the treatment of parameter uncertainty when using PRA results for risk-informed decisionmaking. This guidance addresses the characterization of parameter uncertainty; propagation of uncertainty; assessment of the significance of the state-of-knowledge correlation; (SOKC); and comparison of results with acceptance criteria or guidelines.

- <u>Model Uncertainty</u> – Guidance is provided on how to address the treatment of model uncertainty. This guidance addresses the identification and characterization of model uncertainties in PRAs and involves assessing the impact of model uncertainties on PRA results and insights used to support risk-informed decisions.

The ASME/ANS PRA standard (as endorsed by the NRC) provides requirements that need to be satisfied to understand what sources of uncertainty are associated with a PRA. The guidance developed in this document provides an acceptable approach for meeting the ASME/ANS PRA standard with regard to the requirements on uncertainty (see Section 2.2).

The guidance for the treatment of uncertainties is comprised of seven stages organized into three parts, as illustrated in Figure 1-1.

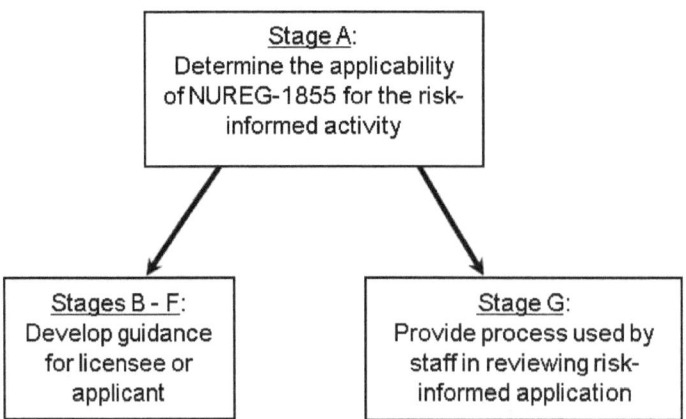

Figure 1-1 Overview of the process stages for the treatment of PRA uncertainties

In Stage A, guidance is provided for assessing the risk-informed activity and associated risk analysis to determine whether the treatment of uncertainties should be based on the approach provided in this NUREG. This guidance generally involves understanding the type of application and the type of risk analysis and results needed to support the application.

In Stages B through F, guidance is provided with regard to the NRC's expectations of a licensee's or applicant's[6] treatment of uncertainties. This guidance generally involves the following:

- Stage B: Understanding risk-informed application and determining the scope of the PRA needed to support the application

- Stage C: Evaluating the completeness uncertainties and determining if bounding analyses are acceptable for the missing scope items

- Stage D: Evaluating the parameter uncertainties

- Stage E: Evaluating model uncertainties to determine their impact on the applicable acceptance guidelines

- Stage F: Developing strategies to address key uncertainties in the application

In Stage G, an overall summary of the process used by the staff is provided with regard to their consideration of uncertainties in their decisionmaking. This process generally involves the following:

- evaluating the PRA for technical adequacy

- determining whether the uncertainties were adequately addressed

---

[6] The process as noted in Section 1.4 supports applications for a design certification or a combined license (COL) to build and operate a reactor. For these types of applications the correct term is "applicant" since a license is not issued for a design certification and has yet to be issued for a COL application. However, for ease of writing, both licensee and applicant will be referred to as a "licensee."

- determining whether the risk element of the risk-informed decisionmaking, in light of the uncertainties, is adequately achieved in the context of the application

- evaluating licensee strategy for addressing the key model uncertainties result in exceeding the acceptance guideline

The approach for addressing the treatment of uncertainties in PRA for risk-informed decisionmaking is summarized in Section 2 and the detailed guidance is provided in Sections 3 through 9.

1.5   Relationship to EPRI Reports on Uncertainties

The NRC staff initiated work on the treatment of uncertainties; however, the NRC recognized from the start that EPRI was also performing work in this area and with similar objectives. Both NRC and EPRI believed a collaborative effort would be more effective and efficient in achieving technical agreement on the subject and would help minimize any duplication of effort. Consequently, NRC and EPRI agreed to work together under a Memorandum of Understanding to ensure the two efforts complemented each other.

In providing guidance on the treatment of uncertainties, both the NRC and EPRI documents start with the specific activity and the decision under consideration to determine whether the treatment of uncertainties should use the approach and guidance provided in NUREG-1855, as complimented by the EPRI reports. In addition, the NRC and EPRI guidance both need to consider (1) the decision under review, (2) the PRA standard, and (3) the supporting PRA model. The NRC approaches the treatment of uncertainties from a regulatory perspective while EPRI approaches the issue from an industry perspective. Both perspectives are essential when using PRA results to support a regulatory decision.

The uncertainties associated with risk contributors modeled in the PRA include the parameter and model uncertainties. Both NRC and EPRI efforts provide guidance for these uncertainties. With regard to parameter uncertainties, this NUREG provides guidance on characterization and propagation while the EPRI reports provide guidance on detailed and approximate methods. With regard to model uncertainties, this NUREG provides guidance on identification of sources of uncertainty that are key to the decision while the EPRI reports provide guidance on the identification and characterization of the uncertainty sources and provides a detailed example of using the process in an application. In addition, this NUREG provides guidance on acceptable bounding analyses for uncertainties related to non-modeled risk contributors. Figure 1-2 below shows the relationship between NRC work (i.e., this NUREG report) and EPRI work.

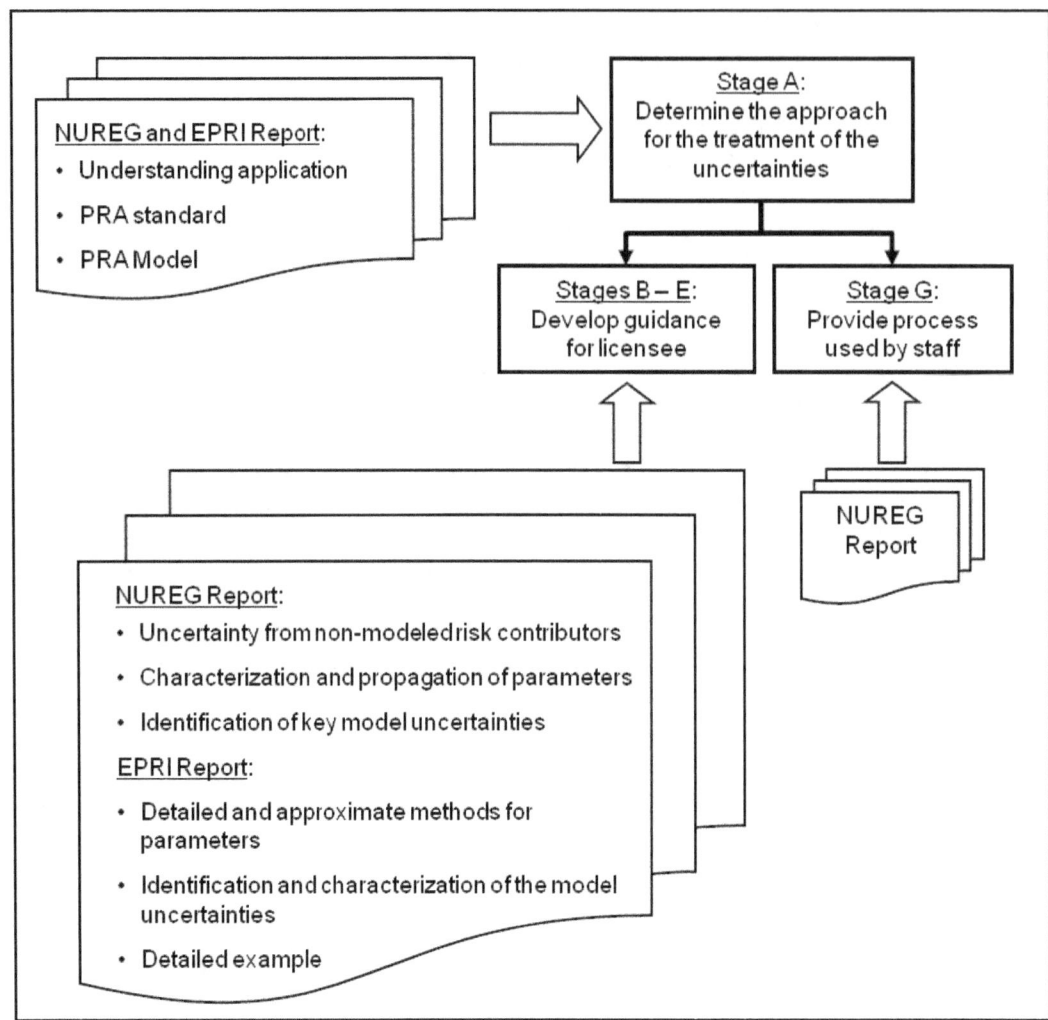

Figure 1-2 Relationship between NRC and EPRI efforts

1.6   Report Organization

The remainder of this report is divided into the following sections:

- Section 2 — Provides an overview of the overall approach used to address uncertainties in risk-informed decisionmaking. This section serves as a roadmap to the rest of the report.

- Section 3 — Provides guidance for Stage A which addresses whether the treatment of uncertainty should use the process in this NUREG

- Section 4 — Provides guidance for Stage B which involves guidance to the licensee for determining if PRA scope and level of detail is adequate to support the application.

- Section 5 — Provides guidance for Stage C which involves guidance to the licensee for addressing the treatment of completeness uncertainty.

- Section 6 — Provides guidance for Stage D which involves guidance to the licensee for addressing the treatment of parametric uncertainty.

- Section 7 — Provides guidance for Stage E which involves guidance to the licensee for addressing the treatment of model uncertainty.

- Section 8 — Provides guidance for Stage F which involves guidance to the licensee for developing a strategy to address the key model uncertainties.

- Section 9 — Provides guidance for Stage G which describes the staff process on addressing the uncertainty in PRA results in the context of risk-informed decisionmaking.

- Section 10 — Provides the references.

2. OVERALL APPROACH

This section provides an overview of the information and guidance provided in the subsequent sections of this report on addressing the treatment of uncertainties associated with probabilistic risk assessment (PRA). The process of addressing the treatment of uncertainties involves the following three phases:

1. determine the approach to use in the treatment of the uncertainties (both the licensee and staff)

2. identify and assess the uncertainties identified by the licensee

3. perform a staff review as part of the risk-informed decisionmaking process

The overall process of addressing the treatment of uncertainties is basically the same regardless of whether the treatment is being addressed by the licensee or applicant or by the staff. As such, both the licensee and the staff need to have a clear understanding of the following

*   the application and the risk contributors that can affect the decision

*   the uncertainties, in the context of the decision under consideration

*   the impact of the uncertainties on the risk results and acceptance guidelines being used to support the decision under consideration

One important interface is the ASME/ANS PRA standard which includes requirements on uncertainties. However, the standard does not provide guidance on how to treat the uncertainties in a PRA, but rather only states that they need to be addressed. This NUREG (in association with Electric Power Research Institute (EPRI) report 1016737, "Treatment of Parameter and Model Uncertainty for probabilistic Risk Assessments" [EPRI, 2008] and EPRI 1026511, "Practical Guidance on the Use of PRA in Risk-Informed Applications with a Focus on the Treatment of Uncertainty" [EPRI, 2012] provide guidance on how to meet the PRA standard's requirements for uncertainties.

The goal of the staff review is to determine whether the licensee has met the risk element of the NRC's risk-informed decisionmaking process (i.e., the decision represents an acceptable risk impact). It is equally important that both the staff and the licensee understand this process. This section provides an overview of what it means to meet the risk element of the NRC's risk-informed decisionmaking process and discusses the following:

*   types of uncertainty (Section 2.1)
*   PRA standard uncertainty requirements (Section 2.2)
*   risk-informed decisionmaking process (Section 2.3)
*   overview of assessing the impact of the uncertainties (Section 2.4)

2.1 Types of Uncertainty

PRA models used to address the risk from nuclear power plants (NPPs) are complex models, the development of which involves a number of different tasks. These tasks include the development of logic structures (e.g., event trees and fault trees) and the assessment of the frequencies and probabilities of the basic events of the logic structures. The development of the logic models and assessment of frequencies and probabilities can introduce uncertainties that could have a significant impact on the results of the PRA model, and these uncertainties need to be addressed. Although uncertainties in a PRA model have different sources, the two basic classes of uncertainties are aleatory and epistemic.

- Aleatory uncertainty is the uncertainty associated with the random nature of events such as initiating events and component failures. PRA models are constructed as probabilistic models and reflect the random nature of the constituent basic events such as the initiating events and component failures. Therefore, a PRA is a probabilistic model that characterizes the aleatory uncertainty associated with accidents at NPPs.

- Epistemic uncertainties arise when making statistical inferences from data and, perhaps more significantly, from incompleteness in the collective state of knowledge about how to represent plant behavior in the PRA model. The epistemic uncertainties relate to the degree of belief that the analysts possess regarding the representativeness or validity of the PRA model and in its predictions (e.g., how well the PRA model reflects the design and operation of the plant and, therefore, how well it predicts the response of the plant to postulated accidents).

Epistemic uncertainties in the PRA models arise for many different reasons, including the following:

- Generally accepted probability models exist for many of the basic events of the PRA model. These models are typically simple mathematical models with only one or two parameters. Examples include the simple constant failure rate reliability model, which assumes that the failures of components in a standby state occur at a constant rate, and the uniformly distributed (in time) likelihood of an initiating event. The model for both these processes is the well-known Poisson model. The parameter(s) of such models may be estimated using appropriate data, which, in the example above, may be comprised of the number of failures observed in a population of like components in a given period of time or the number of occurrences of a particular scenario, in a given period of time, respectively. Statistical uncertainties are associated with the estimates of the model's parameters. Because most of the events that constitute the building blocks of the risk model (e.g., some initiating events, operator errors, and equipment failures) are relatively rare, the data are scarce and the uncertainties can be relatively significant.

- For some events, while the basic probability model is generally accepted, there may be uncertainties associated with the interpretation of the data to be used for estimation of the parameter. For example, when collecting data on component failures from maintenance records, it is not always clear whether the failure would have prevented the component from performing the mission required of it to meet the success criteria assumed in the risk model.

- For some basic events, uncertainty can exist as to how to model the failures, which results in uncertainties in the probabilities of those failures. One example is the behavior of reactor coolant pump (RCP) seals in a pressurized-water reactor (PWR) on loss of cooling. Another example is the modeling of human performance and the estimation of the probabilities of human failure events.

- Uncertainty can exist with regard to a system's ability to perform its function under certain environmental conditions, which are expected to arise during accident scenarios developed in the PRA. This leads to uncertainty in characterizing the success criteria for those functions, which has an impact on the logic structure of the system model. One example is the uncertainty associated with the successful operation of components in the same room of a NPP after a loss of cooling to the room.

As seen in these examples, the uncertainty associated with the structure of and input to the PRA model can be affected by (1) the choice of the logic structure, (2) the mathematical form of the models used to calculate basic event probabilities and frequencies, (3) the model parameter values, or (4) both the mathematical form of the models and the model parameter values together. To the extent that changes in parameter values are little more than subtle changes in the form of the model, it can be argued that no precise distinction exists between model uncertainty and parameter uncertainty. However, as discussed below, parameter uncertainties and model uncertainties are treated differently. Incidentally, it should be noted that, while the Poisson and binomial models are typically adopted for the occurrence of initiating events and for equipment failures, using these models may not be appropriate for all situations.

Epistemic uncertainties are categorized into the following three types: completeness uncertainty, parameter uncertainty, and model uncertainty. The identification, understanding, and treatment of these three types of epistemic uncertainties are the principal subject of the remainder of this report.

2.1.1 Completeness Uncertainty

Completeness uncertainty relates to risk contributors that are not accounted for in the PRA model. This type of uncertainty may further be categorized as either being known, but not included in the PRA model, or unknown. Both known and unknown types of uncertainty are important.

The known completeness uncertainties could have a significant impact on the predictions of the PRA. Examples of sources of these types of incompleteness include the following:

- The scope of the PRA does not include some classes of initiating events, hazards, modes of operation, or component failure modes. That is, some contributors or effects may be knowingly left out of the model for a number of reasons. For example, methods of analysis have not been developed for some issues, and these gaps have to be accepted as potential limitations of the technology. Thus, for example, the impact on actual plant risk from unanalyzed issues cannot now be explicitly assessed. As an additional example, a plant may not currently have a seismic PRA or may have chosen not to model accidents during shutdown modes of operation. Also, the level of detail of the PRA may be limited for some reason, such as cases where medium break loss-of-coolant accidents (LOCAs) are not explicitly included in a PRA but bounded by treating the medium break LOCA as a large break LOCA.

- The level of analysis may have omitted phenomena, failure mechanisms, or other factors because their relative contribution is believed to be negligible. For example, the resources to develop a complete model may be limited, which could lead to a decision not to model certain contributors to risk (e.g., seismically induced fires).

- Some phenomena or failure mechanisms may be omitted because their potential existence has not been recognized or no agreement exists on how a PRA should address certain effects, such as the effects on risk resulting from ageing or organizational factors. Furthermore, PRAs typically do not address them.

Lack of completeness is not in and of itself an uncertainty, but is more of an expression of the limitations in the scope of the model. However, limitations in scope can result in uncertainty about the full spectrum of risk contributors. When a PRA is used to support an application, its scope and level of detail needs to be examined to determine if they match the scope and level of detail required for the risk-informed application. If the scope or level of detail of the existing base PRA is incomplete, then the PRA is either upgraded to include the missing piece(s) or conservative or bounding-type analyses are used to demonstrate that the missing elements are not significant risk contributors.

The guidance in Section 5 focuses on the use of conservative- and bounding-types of analyses to address the treatment of completeness uncertainty. However, this approach can only be used for those sources of completeness uncertainty that are known to exist. Unknown sources of completeness uncertainty are addressed in risk-informed decisionmaking by other methods, such as safety margins, as discussed in Section 9.

In the context of an application, the scope and level-of-detail items that are outside of the needed scope of the PRA or have a greater level of detail than is needed, respectively, may be addressed by expanding the scope and level of detail to include those items or by performing a bounding analysis to demonstrate that those items are not significant risk contributors.

2.1.2 Parameter Uncertainty

Parameter uncertainty relates to the uncertainty in the computation of the input parameter values used to quantify the frequencies and probabilities of the events in the PRA logic model. Examples of such parameters are initiating event frequencies, component failure rates and probabilities, and human error probabilities. These uncertainties can be characterized by probability distributions that relate to the analysts' degree of belief in the values of these parameters (which could be derived from simple statistical models or from more sophisticated models).

As part of the risk-informed decisionmaking process, the numerical results (e.g., CDF and LERF) of the PRA, including their associated quantitative uncertainty, are compared with the appropriate acceptance guidelines. The uncertainties on the input parameters need to be propagated through the risk calculations in an appropriate manner to provide an assessment of the quantitative uncertainty on the PRA results. An important aspect of this propagation is the need to account for what is known as the state-of-knowledge correlation (SOKC).

For many parameters (e.g., initiating event frequencies, component failure probabilities or failure rates, human error probabilities), the uncertainty is characterized as probability distributions that represent the degree of belief in the value of the parameter. Section 6 discusses the methods for propagating these uncertainties through the PRA model to characterize the uncertainty in the numerical results of the analysis. In this manner, the impact of the parameter uncertainties on the numerical results of the PRA can be assessed integrally. However, many of the acceptance criteria or guidelines used in risk-informed decisionmaking (e.g., the acceptance guidelines of Regulatory Guide (RG) 1.174) are defined such that the appropriate measure for comparison is the mean value of the uncertainty distribution on the corresponding metric. In this case, as discussed in Section 6, the primary issue with parameter uncertainty is its effect on the calculation of the mean, and specifically, on the relevance and significance of the SOKC.

The SOKC is important when the same data is used to quantify the individual probabilities of two or more basic events. The uncertainty associated with such basic event probabilities must be correlated to correctly propagate the parameter uncertainty through the risk calculation. Most PRA software in current use has the capability to propagate parameter uncertainty through the analysis while taking into account the SOKC to calculate the probability distribution for the results of the PRA. In some cases, however, it may not be necessary to consider the SOKC. Section 6 examines the implications of and provides guidance on when it is important to account for the SOKC.

> EPRI report 1016737 [EPRI, 2008] provides guidance for ascertaining the importance of the SOKC.

2.1.3 Model Uncertainty

Model uncertainty relates to the uncertainty associated with some aspect of a PRA model that can be represented by any one of several different modeling approaches, none of which is clearly more correct than another. Consequently, uncertainty is introduced into the PRA results since there is no consensus about which model most appropriately represents the particular aspect of the plant being modeled.

Model uncertainty is related to an issue for which no consensus approach or model exists and where the choice of approach or model is known to have an effect on the PRA model (e.g., the introduction of a new basic event, changes to basic event probabilities, change in success criterion, and the introduction of a new initiating event). Model uncertainty may result from a lack of knowledge about how structures, systems and components (SSCs) behave under the conditions that arise during the development of an accident. A model uncertainty can arise for the following reasons:

- The phenomenon being modeled is itself not completely understood (e.g., behavior of gravity-driven passive systems in new reactors, or crack growth resulting from previously unknown mechanisms).

- For some phenomena, other data or information may exist, but needs to be interpreted to infer SSC behavior under conditions different from those in which the data were collected (e.g., RCP seal LOCA information).

- The nature of the failure modes is not completely understood or is unknown (e.g., digital instrumentation and controls).

Another source of model uncertainty is the uncertainty associated with the logic structure of the PRA model or in the choice of model used to estimate the frequencies or probabilities associated with the basic events, or both.

The uncertainty associated with a model and its constituent parts is typically addressed by making assumptions. Examples of such assumptions include those concerning (1) how a reactor coolant pump in a PWR would fail following loss-of-seal cooling, (2) the approach used to address common cause failure in the PRA model, and (3) the approach used to identify and quantify operator errors. In general, model uncertainties are addressed by determining the sensitivity of the PRA results to different assumptions or models.

The treatment of model uncertainty in risk-informed decisionmaking depends on how the PRA will be used. PRAs can be used in two ways:

1.   The results of the base PRA[7] are used as the input for the evaluation of the significance of temporary or permanent changes to a plant's licensing basis (e.g., by using RG 1.174) with emphasis on the change in risk due to the proposed plant changes (design or operational).

2.   The results of the base PRA are used in a regulatory application (e.g., to evaluate various design options or to determine the baseline risk profile as part of a license submittal for a new plant).

The expectation is that the focus in the decisionmaking will be on identifying and evaluating sources of model uncertainty and related assumptions that are key to the specific application at hand. Identifying the key sources of model uncertainties and related assumptions involves the following three steps:

1.   <u>Identification of Sources of Model Uncertainties and Related Assumptions of the Base PRA</u> – Both generic and plant-specific sources of model uncertainty and related assumptions for the base PRA are identified and characterized. These sources of uncertainty and related assumptions are those that result from developing the PRA model.

2.   <u>Identification of Sources of Model Uncertainties and Related Assumptions Relevant to the Application</u> – The sources of model uncertainty and related assumptions in the base PRA that are relevant to the application are identified. This identification may be performed with a qualitative analysis. This analysis is based on an understanding of how the PRA is used to support the application and the associated acceptance criteria or guidelines. In addition, new sources of model uncertainty and related assumptions that may be introduced by the application are identified.

3.   <u>Identification of Key Sources of Model Uncertainties and Related Assumptions for the Application</u> – The sources of model uncertainty and related assumptions that are key to the application are identified. Quantitative analyses of the importance of the sources of model uncertainty and related assumptions identified in the previous steps are performed in the context of the acceptance guidelines for the application. The analyses are used to identify any credible alternative modeling hypotheses that could impact the decision. These hypotheses are used to identify which of the sources of model uncertainty and related assumptions are key to the application.

Section 7 provides detailed guidance on the treatment of model uncertainty.

Although parameter uncertainty analyses methods are fairly mature and are addressed adequately through the use of probability distributions on the values of the parameters, the analysis of the model and completeness uncertainties is typically not handled in such a formal

---

[7] The term "base PRA" is meant to denote the PRA that is developed to support the various applications; that is, the base PRA is independent of an application.

manner. The typical response to a modeling uncertainty is to choose a specific modeling approach to be used in developing the PRA model. Although it is possible to embed a characterization of model uncertainty into the PRA model by including several alternate models and providing weights (probabilities) to represent the degree of credibility of the individual models, this approach is not typical. Notable exceptions are NUREG-1150 [NRC, 1990] and the approach to seismic hazard evaluation proposed by the Senior Seismic Hazard Analysis Committee [LLNL, 1997]. The approach taken in this document is, the following: when using the results of the PRA model, it is necessary to determine whether uncertainties, credible alternative hypotheses, or choice of modeling methods would significantly change the assessment. Section 7 discusses methods for performing such demonstrations.

In dealing with model uncertainties, it is helpful to identify whether the model uncertainties can alter the logic structure of the PRA model or whether the model uncertainties primarily impact the frequencies or probabilities of the basic events of the logic model, or both. For example, an uncertainty associated with the establishment of the success criterion for a specific system can result in an uncertainty regarding whether one or two pumps are required for a particular scenario. In this example, the uncertainty would be reflected in the choice of the top gate in the fault tree for that system. On the other hand, an example of model uncertainties that do not alter the structure of the model are those model uncertainties associated with the choice of model used to estimate the probabilities of the basic events. Tools, such as importance analyses, can be used to explore the potential impact of this type of model uncertainty in a way not possible for those uncertainties related to the logic structure of the PRA model.

One approach to dealing with a specific model uncertainty is to adopt a consensus model that essentially eliminates the need to address the model uncertainty. In the context of regulatory decisionmaking, a consensus model can be defined as follows:

> **Consensus model** – In the most general sense, a consensus model is a model that has a publicly available published basis[8] and has been peer reviewed and widely adopted by an appropriate stakeholder group. In addition, widely accepted PRA practices may be regarded as consensus models. Examples of the latter include the use of the constant probability of failure on demand model for standby components and the Poisson model for initiating events. For risk-informed regulatory decisions, the consensus model approach is one that NRC has utilized or accepted for the specific risk-informed application for which it is proposed.

The definition given here ties the consensus model to a specific application. This restriction exists because models have limitations that may be acceptable for some uses and not for others. In some cases (e.g., the Westinghouse Owners' Group 2000 RCP seal LOCA model), this consensus is documented in a safety evaluation report (SER) [NRC, 2003d]. In many cases, the model tends to be considered somewhat conservative. In this case, it is important to recognize the potential for this conservatism to mask other contributors that may be important to a decision. Many models can already be considered consensus models without the issuance of an SER. For example, the Poisson model for initiating events has been used since the very early days of PRA.

---

[8] It is anticipated that most consensus models would be available in the open literature. However, under the requirements of 10 CFR 2.390, there may be a compelling reason, for exempting a consensus model from public disclosure.

There may be cases where there may be more than one consensus model for addressing a specific issue. An example is the Multiple Greek Letter and the Alpha methods for quantifying common cause failures. In such a case, any one of the consensus models can be used. Multiple consensus models should provide similar results. If they do not, then they do not meet the requirement for being a consensus model and an evaluation of the associated model uncertainty should be made utilizing the guidance in Section 7. It should also be noted that adoption of a consensus model would not negate the need to model any parameter uncertainty associated with the consensus model. An example of this situation is discussed in section 4 of a paper by Zio and Apostolakis [Zio, 1996]. Adoption of consensus models obviates the need to consider other models as alternatives.

Currently there is no agreed-on list of consensus models, however, as a first step in establishing such a process, EPRI has compiled a list of candidate consensus models [EPRI, 2006a]. This list includes common approaches, models, and sources of data used in PRAs. At this time, the NRC has not reviewed this list although specific models, approaches and data may have been approved for specific risk-informed applications.

> EPRI reports 1016737 [EPRI, 2008] and 1026511 [EPRI, 2012] provide a generally acceptable approach for identifying sources of model uncertainties and related assumptions; it also provides a generic list of sources of model uncertainties and

2.2 PRA Standard Uncertainty Requirements

The ASME/ANS PRA standard provides requirements (i.e., supporting requirements [SRs]) for addressing both parameter and model uncertainties. The SRs are written in terms of capability categories (CCs) that define the extent to which different aspects of the PRA model may vary. This variation is expressed in terms of (1) the extent to which the scope and level of detail of plant design, operational and maintenance are modeled, (2) the extent to which plant specific information on equipment performance is included, and (3) the extent to which realism of the plant response is addressed. There are a total of three CCs (written as CC I, CC II, or CC III), each of which may have its own requirement; however, some requirements extend across two or more of these categories, as appropriate.

In the standard, parameter uncertainties are associated with the calculation of the following:

- initiating event frequencies

- basic event failure probabilities

- human error probabilities

- risk metrics such as core damage frequency (CDF) and large early release frequency (LERF)

- Other PRA inputs (e.g., hazard intensity, plant fragility)

The standard (as endorsed by the NRC) requires the calculation of mean values for the parameters which are used to calculate the either the frequency or probability of the significant contributors. The standard also requires that their probabilistic representation of the uncertainty be provided for these quantities.

For CDF and LERF, the standard (with NRC endorsement) requires that a mean value be calculated that is based on the mean values of the significant input parameters and that (1) the state-of-knowledge correlation between significant event frequencies or probabilities is taken into account, and (2) if the state-of-knowledge correlation between event frequencies or probabilities is significant, then propagate the uncertainty distribution for those parameters when calculating CDF or LERF.

2.3 The Risk-Informed Decisionmaking Process

In a white paper, "Risk-Informed and Performance-Based Regulation" [NRC, 1999], the Commission defined a risk-informed approach to regulatory decisionmaking:

> A "risk-informed" approach to regulatory decisionmaking represents a philosophy whereby risk insights are considered together with other factors to establish requirements that better focus licensee and regulatory attention on design and operational issues commensurate with their importance to public health and safety.

This philosophy was elaborated on in RG 1.174 [NRC, 2002a] to develop a risk-informed decisionmaking process for licensing changes and has since been implemented in NRC risk-informed activities.

In developing the risk-informed decisionmaking process, the NRC defined a set of key principles in RG 1.174 to be followed for risk-informed decisions regarding plant-specific changes to the licensing basis. The following principles are global in nature and can be generalized to all activities that are the subject of risk-informed decisionmaking:

- Principle 1: Current Regulations Met
- Principle 2: Consistency with Defense-in-Depth Philosophy
- Principle 3: Maintenance of Safety Margins
- Principle 4: Acceptable Risk Impact
- Principle 5: Monitor Performance

The principles of risk-informed decisionmaking are expected to be observed; however, they do not describe the process that is used in risk-informed decisionmaking. RG 1.174 presents an approach that ensures the principles will be met for risk-informed decisionmaking involving plant-specific changes to the licensing basis. This approach can be generalized and applied to all risk-informed decisionmaking.

The generalized approach integrates all the insights and requirements that relate to the safety or regulatory issue of concern. These insights and requirements include recognition of any mandatory requirements resulting from current regulations as well as the insights from deterministic and probabilistic analyses performed to help make the decision. The generalized approach ensures that defense-in-depth measures and safety margins are maintained. It also includes provisions for implementing the decision and for monitoring the results of the decision. Figure 2-1 provides an illustration of this integrated process.

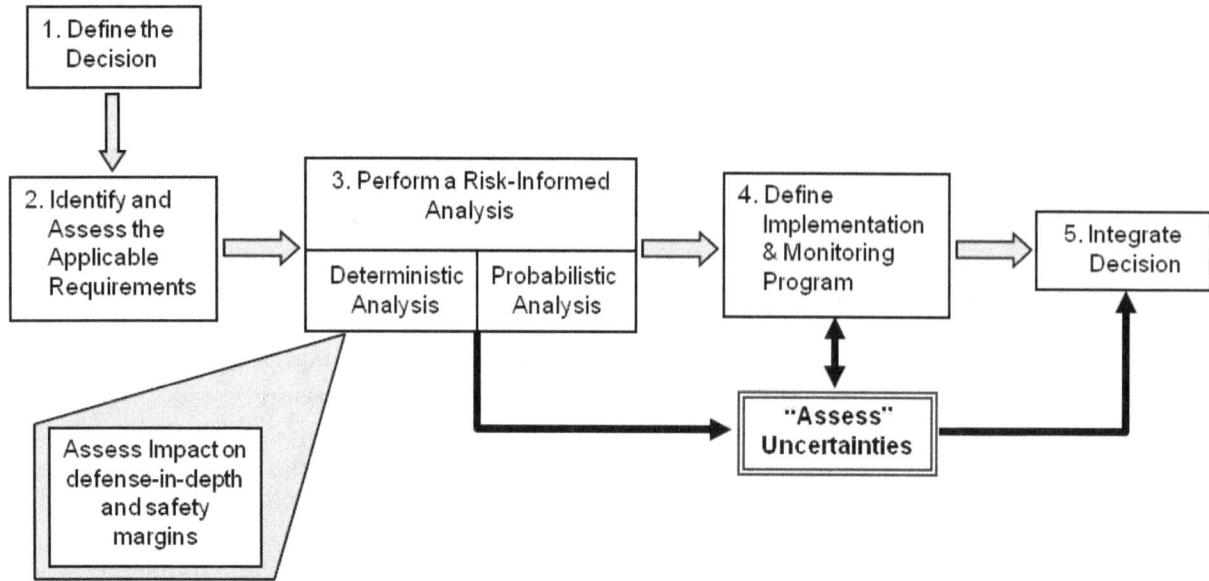

Figure 2-1 Elements of the integrated risk-informed decisionmaking process

Element 1: Define the decision. The first step in the process is to define the issue or decision under consideration. Some examples of the types of issues/decisions that NRC would need to address are related to:

- the design or operation of the plant

- the plant technical specifications/limits and conditions for normal operation

- the periodicity of in-service inspection, in-service testing, maintenance, and planned outages

- the allowed combinations of safety system equipment that can be removed from service during power operation and shutdown modes

- the adequacy of the emergency operating procedures and accident management methods

Element 2: Identify and assess the applicable requirements. In this element, the current regulatory requirements that apply to the decision under consideration are identified. Part of this determination includes identifying (understanding) the effect of the applicable requirements on the decision. This element implements Principle 1 of risk-informed decisionmaking.

Element 3: Perform a risk-informed analysis. In this element, an assessment is made, in terms of a risk-informed analysis, to demonstrate that Principles 2, 3, and 4 are met. The risk-informed analysis includes both deterministic and probabilistic components. Appropriate consideration of the uncertainty in both deterministic and probabilistic assessments is required to properly interpret the results. Both the deterministic and probabilistic components implement Principles 2 and 3, which take into account the impact on defense-in-depth and on safety margins. The probabilistic component implements Principle 4, acceptable risk impact. A treatment of the uncertainties in the probabilistic analysis is implicitly required to implement Principles 2, 3, and 4 of risk-informed decisionmaking. Treatment of probabilistic analysis uncertainties is the focus of this report. Although uncertainties in a deterministic analysis are not explicitly addressed in this report, the types of uncertainties and the methods for evaluating them are the same for a deterministic assessment.

Element 4: Define Implementation and Monitoring Program. A part of the decisionmaking process involves understanding the effect of implementing a positive decision. This understanding involves determining how to monitor the change so that future assessments can be made as to whether the decision was implemented effectively and to guard against any unanticipated adverse effects. Consequently, consideration should be given to a performance-based means of monitoring the results of the decision.

Element 5: Integrated decision. In this final element of the decisionmaking process, the results from Elements 1 through 4 are integrated and it is decided whether to accept or reject the application. This integration requires that the individual insights obtained from the other elements of the decisionmaking process are weighed and combined to reach a conclusion. An essential aspect of the integration is the consideration of uncertainties.

PRAs can address many uncertainties explicitly. These uncertainties are the epistemic uncertainties arising from recognized limitations in knowledge. However, a specific type of uncertainty exists that risk analyses, whether deterministic or probabilistic, cannot address. This type of uncertainty relates to the lack of a complete state of knowledge about potential failure modes or mechanisms. Because these failure modes or mechanisms are unknown, they cannot be addressed analytically (whether the analysis is deterministic or probabilistic). Principles 2, 3, and 5 (i.e., those related to defense-in-depth, safety margins, and performance monitoring) address the unknown type of completeness uncertainty. The guidance in this report focuses on the treatment of the parameter and model uncertainties associated with the PRA and, in particular, those of known completeness uncertainties that are not modeled in the PRA.

The above five elements constitute the steps of an integrated risk-informed decisionmaking process. However, in providing guidance on the treatment of uncertainties, it is important to understand the implementation of these elements by the licensee and by the NRC staff. While the licensee and the NRC both have the same goal, the actual implementation of the process by each entity is different. The licensee identifies the uncertainties and determines their impact on the PRA results and, ultimately, on the acceptance guidelines of the decision under consideration. The NRC staff is determines whether the process followed by the licensee is adequate.

The guidance for determining the impact of uncertainties on a risk-informed decisionmaking is summarized below in Section 2.4.

2.4   Assessing the Impact of the Uncertainties

As discussed previously, the guidance for treating PRA uncertainties in the risk-informed decisionmaking process are organized into three parts: (1) determining the approach to use in the treatment of the uncertainties (both the licensee and NRC); (2) identifying and assessing the uncertainties (performed by the licensee); and (3) describing the staff review process as part of the risk-informed decisionmaking process. These three parts are comprised of seven stages, as illustrated in Figure 2-2.

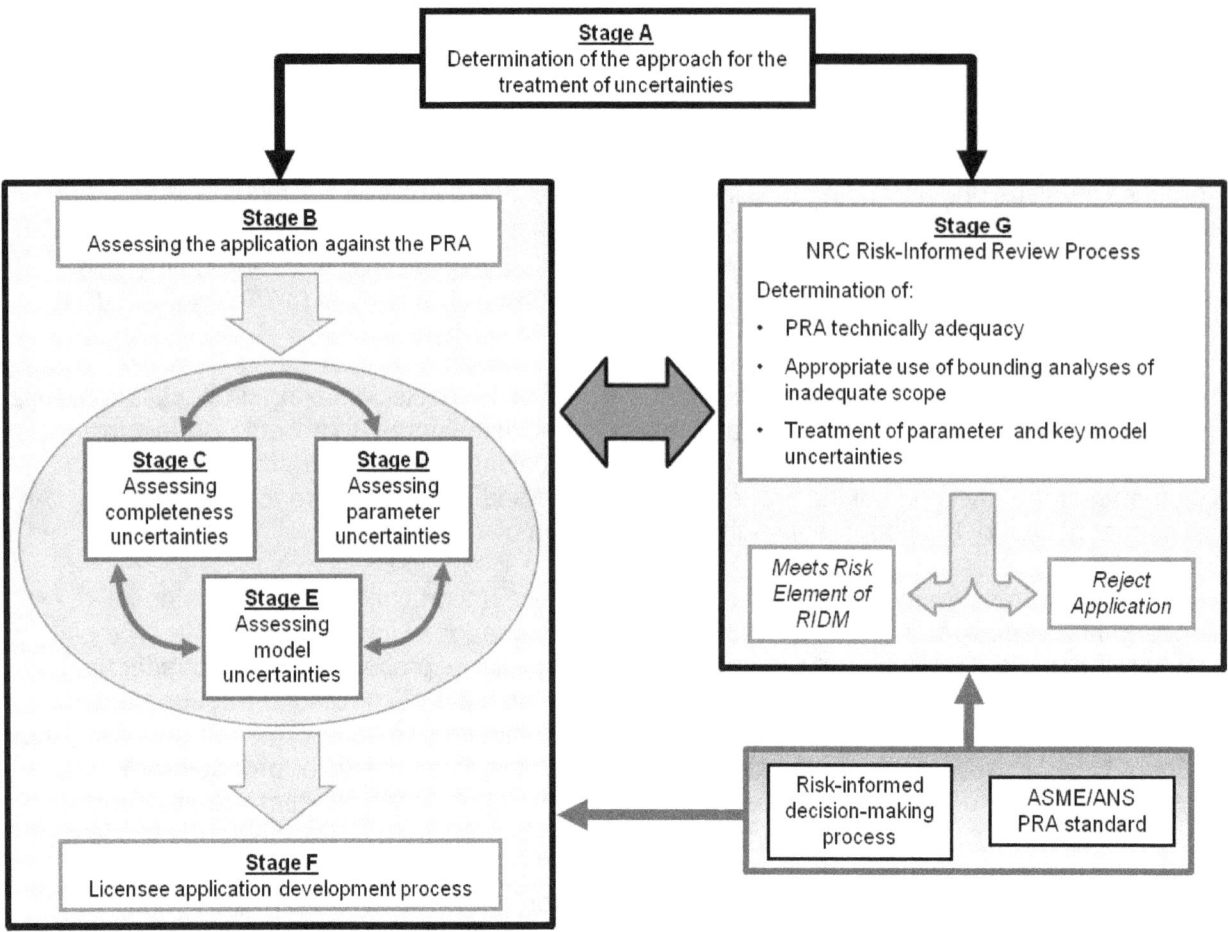

Figure 2-2 Process for the PRA treatment of uncertainties in risk-informed decisionmaking

Stage A: Determining the Approach for the Treatment of Risk Uncertainties

This stage provides guidance to both the licensee and the NRC on determining whether the activity under consideration fits within the scope of the guidance provided in this NUREG. That is, it is necessary for both the NRC and the licensee to be in agreement whether the treatment of uncertainties are to be addressed using the guidance provided in this NUREG. It is recognized that, depending on the application (i.e., the decision under consideration), the treatment of uncertainties can vary. Consequently, although the uncertainties always need to be addressed, how the uncertainties are addressed in the decision is dependent on the activity and the type of risk results being used.

Stage B: Assessing PRA Scope and Level of Detail

This stage provides guidance to the licensee on determining whether their PRA can support the application. In this stage, the treatment of uncertainties involves (1) assessing the application to determine the PRA scope and level of detail needed to support the application, and then (2) determining if the actual PRA achieves the needed scope and level of detail. If the scope and level of detail of the PRA is inadequate for the application, the licensee may choose to redefine the application such that the scope and level of detail of the existing PRA is adequate to support the application; upgrade the PRA to include the missing scope or level of detail; or demonstrate that the missing scope or level of detail is not significant to the decision.

Stage C: Assessing Completeness Uncertainties

This stage provides guidance to the licensee on assessing the completeness uncertainty. This stage is only invoked if the PRA scope or level of detail has been determined to be inadequate and the licensee has decided neither to redefine the application nor to upgrade the PRA. At this stage of the process, the licensee is performing analyses to determine the risk significance of the non-modeled items in the PRA. That is, the licensee has decided to assess the significance of the lack of completeness. If the non-modeled items are risk significant, the licensee must still address the lack of completeness of the PRA. The licensee can choose to redefine the application such that the scope and level of detail of the existing PRA is adequate to support the application; upgrade the PRA to include the missing scope or level of detail; or address the lack of completeness via some deterministic means.

Stage D: Assessing Parameter Uncertainties

This stage provides guidance to the licensee on assessing the parameter uncertainties and determining their impact on the application-specific acceptance guidelines. As noted earlier, the PRA standard provides requirements for parameter uncertainties; however, the standard does not prescribe an approach for how to assess the uncertainties. The standard only states that uncertainties are to be assessed. The guidance in this NUREG provides an acceptable approach on the characterization and propagation of the parameter uncertainties. However, the EPRI report [EPRI, 2008] provides guidance for determining the importance of the SOKC in carrying out the propagation. Once the parameter uncertainties have been assessed, guidance is provided for their treatment when determining whether the quantitative acceptance guidelines have been challenged. This step is an important element in the risk-informed decisionmaking process since the uncertainties may be responsible for challenging the acceptance guidelines or may further demonstrate that the acceptance guidelines cannot be met.

Although acceptance guidelines may not be challenged at this stage of the process, they may be challenged when model uncertainties are factored into risk estimate calculations. Consequently, the impact of the model uncertainties must also be assessed.

Stage E: Assessing Model Uncertainties

This stage provides guidance to the licensee on identifying the model uncertainties and determining their impact on the application-specific acceptance guidelines. Similar to parameter uncertainties, the PRA standard also provides requirements for model uncertainties; however, the standard only requires that they be identified and characterized and that their effect upon the PRA model is identified. The standard does not include guidance on how to characterize or assess the model uncertainties, or how to identify their impact on the PRA model; the guidance in this NUREG provides an acceptable approach. Furthermore, while every model uncertainty needs to be identified, not every model uncertainty is relevant to the decision under consideration. Guidance is provided to the licensee in this stage on making this determination. As with the parameter uncertainties, the model uncertainties may be responsible for challenging the acceptance guidelines or may further demonstrate that the acceptance guidelines cannot be met.

Stage F: Licensee Application Development Process

This stage provides guidance to the licensee on determining the strategy for addressing the key uncertainties that contribute to risk metric calculations that challenge application-specific acceptance guidelines. The licensee may need to adjust the application for one or more of the following reasons:

- The scope of the PRA is incomplete.

- The results from the PRA may be challenging the acceptance guidelines with or without the parameter uncertainties.

- The results from the PRA may be challenging the acceptance guidelines with or without parameter uncertainties and with or without model uncertainties.

Under these circumstances, the licensee can upgrade or refine the PRA, refine the application, or address the impact of the uncertainties via other means (e.g., deterministic analyses). In addition, the licensee may choose to implement compensatory measures.

At this point in the process, the guidance changes focus to the NRC staff review. However, this guidance describes how the NRC staff review determines whether the risk analysis element of the risk-informed decisionmaking process has been met and whether the uncertainties were adequately addressed.

Stage G: NRC Risk-Informed Review Process

The staff review of a risk-informed application begins with the comparison of the application risk results to the acceptance guidelines. The justification needed to demonstrate the acceptability of a given risk-informed application is largely dictated by the proximity of the risk results to the acceptance guidelines. In general, an application can be characterized as falling into one of the following four general regimes based on the proximity of the risk results to the acceptance guidelines:

- Regime 1—The risk results are well below the acceptance guidelines
- Regime 2—The risk results are closer to, but do not challenge the acceptance guidelines
- Regime 3—The risk results challenge the acceptance guidelines
- Regime 4—The risk results clearly exceed the acceptance guidelines

The justification for a given application should be commensurate with the proximity of the risk results to the acceptance guidelines, as shown above. In general, more justification will be needed for a given application when the risk results are closer to challenging or exceeding the acceptance guidelines than when the risk results are further away.

In determining whether the acceptance guidelines have been met, the staff seeks to answer the following general questions:

- How do the risk results compare to the acceptance guidelines?
- Is the scope and level of detail of the PRA appropriate for the application?
- Is the PRA model technically adequate?
- Is the acceptability of the application adequately justified?

Similar to the licensee's development of the risk-informed application, the staff's risk-informed review process is not necessarily performed in a serial manner, nor is the transition from one portion of the review process to another always absolute. The staff's risk-informed review is a dynamic process that often relies on additional information beyond the original application submittal that the NRC may request from the licensee. In general, when the staff makes a determination in a given step of the process, the determination is based on a review of the submittal documentation in conjunction with any information received via open and continuous dialogue with the licensee. This dialogue is meant to achieve the clearest understanding of the application and generally consists of oral discussions and written correspondence. This dialogue may also result in the generation of official requests for additional information by the staff, all of which are formally documented and considered together with the original submittal. In this way, the staff considers the original submittal documentation and any additional information and insights gained from the review process as a whole during the risk-informed review of the licensee's application.

3. STAGE A — THE APPROACH FOR TREATING RISK ANALYSIS UNCERTAINTIES

This section provides guidance to both the licensee and the Nuclear Regulatory Commission (NRC) staff on determining whether the approach for treating probabilistic risk assessment (PRA) uncertainties, as provided in this NUREG, should be used for the risk-informed activity (i.e., the decision) under consideration. This guidance is provided to both the licensee and the staff as it is important for both the NRC and the licensee to be in agreement on whether the guidance provided in this NUREG should be used for the treatment of uncertainties.

Although the process for addressing uncertainties outlined this NUREG is generic in nature, it has been developed specifically to support risk-informed regulatory activities that are licensee-initiated. The generic application of this process for other risk-informed activities is described in Section 3.4.

Although uncertainties always need to be addressed in risk-informed activities, the approach used to address uncertainties can vary and is dependent on the nature of the risk-informed activity under consideration. As such, this guidance discusses two steps that are used to determine whether the approach for the treatment of uncertainties, as described in this NUREG, is applicable for a given risk-informed activity. This guidance involves determining the following:

- The type of risk results used in the application – Are the risk results PRA or non-PRA in nature?

- Application of PRA results – If the results are from a PRA, how are the results being used to support the decision?

This applicability determination process is illustrated in Figure 3-1 and summarized below.

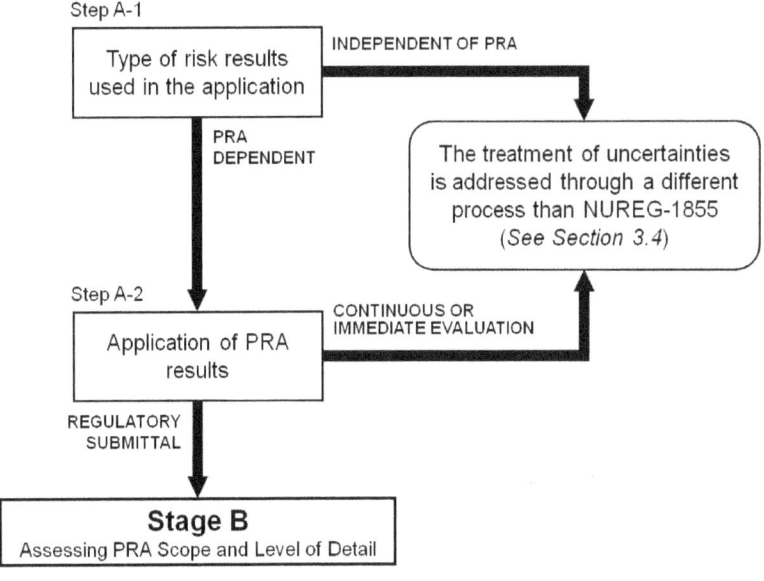

Figure 3-1 Overview of Stage A

As shown above, the applicability determination process described in this Section starts with a determination of the type of risk-informed activity that is being evaluated (non-risk-informed activities are outside the scope of this report) and the type of risk results used in the application. If the risk results do not depend on the development of a PRA, the approach in this NUREG does not apply. Next, with regard to how the PRA is being used in the decision, if the activity involves a continuous evaluation of risk or an evaluation of risk at the time of an event, the approach in this NUREG does not apply. The following sections go into more detail for both steps of the process used to determine whether the approach in this NUREG is applicable to a given risk-informed activity.

3.1   Step A-1: Type of Risk Results

The purpose of this step is to determine whether the risk results used in the decision require the development of a PRA model. In general, a risk analysis may be either quantitative or qualitative and both methods will have associated uncertainties. This NUREG provides an approach (guidance) for the treatment of uncertainties relative to quantitative acceptance guidelines. That is, for many decisions, the primary measure for determining acceptability of the decision is whether the quantitative acceptance guidelines are challenged by the risk results, and if so, the extent to which they are challenged and perhaps exceeded. For a given decision, the uncertainties may affect the risk results such that the acceptance guidelines are challenged or exceeded.

The approach provided in this NUREG has been specifically developed for the treatment of uncertainties associated with a PRA. If a given risk-informed activity is determined to be a licensee-initiated regulatory activity, but does not utilize the results from a PRA, it is not subject to the approach for the treatment of uncertainties provided in this NUREG (See Section 3.4).

The risk analysis being performed is dependent on the risk metrics being evaluated. In general, a PRA is a risk analysis that produces a variety of results that can be important to risk-informed decisionmaking. The following are some examples of these results:

- overall core damage frequency (CDF) or large early release frequency (LERF)
- accident sequence CDF or LERF
- importance measures

Risk metrics, such as overall or individual accident sequence CDF or LERF and importance measures, require that PRA model be developed and solved to produce the quantitative results and are considered outputs of a PRA model.

Initiating event frequencies, component failure probabilities, and human error probabilities are inputs to the PRA model and, although they must necessarily be quantified when performing a PRA, these items can also be quantified separately from a PRA and are therefore independent from the development of a PRA model. A risk-informed activity may use these input parameters instead of evaluating a risk metric.

For example, the proposed alternative ECCS rule (10CFR 50.46 (a))[9] establishes a transition break size (TBS) that delineates where current evaluation methods have to be utilized (i.e., for breaks less than the TBS) and where less conservative methods can be applied (i.e., for breaks greater than the TBS). The computation of pipe break frequencies used to help determine the

---

[9] This alternative regulation has not been approved by the Commission.

TBS does include parameter uncertainties but is not dependent on the development of a PRA model. Consequently, the uncertainties from the PRA are not relevant to this type of activity and the risk results used in this activity do not depend on the development of a PRA. Therefore, the treatment of uncertainties in the parameters being used to support the proposed alternative ECCS rule would not be subject to the approach for the treatment of uncertainties presented in this NUREG.

Table 3-1 lists licensee-initiated risk-informed regulatory activities. This table describes some common risk-informed initiatives and is not meant to be comprehensive. The activities that require the development of a PRA are indicated. Note that some NRC activities such as Notice of Enforcement Discretion and the Significance Determination Process are included in this table since they can include consideration of risk information provided by licensees. Whether the licensee-initiated activities identified in Table 3-1 are ultimately subject to the approach for the treatment of uncertainties described in this NUREG is dependent on how the PRA results are used to support the decision under consideration.

Table 3-1 Candidate risk-informed activities; licensee-initiated regulatory activities.

Risk-informed activities	Type of risk metrics utilized/acceptance guidelines	Is PRA required?
Component Risk Ranking (Motor-Operated Valves)	Utilizes the Fussell-Vesely (FV) importance measure calculated from a PRA to rank components. Importance values are provided for ranking components into High, Medium, and Low categories.	Yes
Mitigating Systems Performance Index (MSPI)	MSPI is the sum of changes in a simplified CDF evaluation resulting from changes in a systems unavailability and unreliability relative to baseline values.	Yes
NFPA 805 Fire Protection 10CFR 50.48(c)	Utilizes the average annual CDF and LERF calculated from the baseline fire PRA. The total change in risk associated with a licensee's transition to NFPA 805 should be consistent with the acceptance guidelines of Regulatory Guide (RG) 1.174.	Yes
Notice of Enforcement Discretion (NOED)	The NRC can exercise enforcement discretion with regard to unanticipated temporary noncompliance with license conditions and Technical Specifications (TS). A licensee may depart from its TS in an emergency, pursuant to the provisions of 10 CFR 50.54(x), without prior NRC approval, when it must act immediately to protect the public health and safety. However, situations occur occasionally that are not addressed by the provisions of 10 CFR 50.54(x), and for which the NRC's exercise of enforcement discretion may be appropriate. The licensee may provide risk input to the NRC to support their decision.	Possibly

Table 3-1 Candidate risk-informed activities; licensee-initiated regulatory activities (Continuation).

Risk-informed activities	Type of risk metrics utilized/acceptance guidelines	Is PRA required?
Regulatory Guide 1.174	Utilizes the average annual CDF and LERF calculated from the baseline PRA and the change in CDF (ΔCDF) or change in LERF (ΔLERF) evaluated with proposed modifications to the plant design or operation reflected in the PRA.	Yes
Regulatory Guide 1.175	For RG 1.175 applications, risk-information is used in two main areas. The first is using PRA determined importance measures (FV and Risk Achievement Worth (RAW)) to establish the safety significance of components and classifying them as either high safety significant or low safety significant components. The second area where risk information is used is in evaluating the risk increase resulting from changes in the IST program. The risk increase must be compatible with the criteria of RG 1.174.	Yes
Regulatory Guide 1.177	For evaluating the risk associated with proposed technical specification allowed outage time changes, a three tiered approach is discussed in the regulatory guide. Tier 1 is an evaluation of the impact on plant risk of the proposed TS change as expressed by ΔCDF, the incremental conditional core damage probability (ICCDP), and, when appropriate, ΔLERF and the incremental conditional large early release probability (ICLERP). Tier 2 is an identification of potentially high-risk configurations that could exist if equipment in addition to that associated with the change were to be taken out of service simultaneously, or other risk significant operational factors such as concurrent system or equipment testing were also involved. The objective of this part of the evaluation is to ensure that appropriate restrictions on dominant risk-significant configurations associated with the change are in place. Tier 3 is the establishment of an overall configuration risk management program to ensure that other potentially lower probability, but nonetheless risk-significant, configurations resulting from maintenance and other operational activities are identified and compensated for.	Yes
Regulatory Guide 1.178	For RG 1.178 applications, risk-information is used to categorize piping segments into high-safety-significant and low-safety-significant classifications, and to estimate the change in risk resulting from a change in the ISI program. The change in risk is evaluated and compared to the guidelines presented in RG 1.174.	Yes

Table 3-1 Candidate risk-informed activities; licensee-initiated regulatory activities (Continuation).

Risk-informed activities	Type of risk metrics utilized/acceptance guidelines	Is PRA required?
Significance Determination Process (SDP)	The SDP uses risk insights, where appropriate, to help NRC inspectors and staff determine the safety or security significance of inspection findings. The safety significance of findings, combined with the results of the performance indicator (PI) program, is used to define a licensee's level of safety performance and to define the level of NRC engagement with the licensee. Risk insights can be developed by the NRC and in some cases, provided by a licensee.	Possibly
Technical Specification Initiative 4b Risk-Informed Completion Times	PRA methods are used to calculate the configuration-specific risk in terms of CDF and LERF. These risk metrics are applied to determine an acceptable extended duration for the completion time (CT). There are two components to the risk impact that is addressed: (1) the single event risk when the CT extension is invoked and the component is out-of-service, and (2) the yearly risk contribution based on the expected frequency that the CT extension will be implemented. The yearly risk impact is represented by the ΔCDF and the ΔLERF metrics referenced in RG 1.174. The single event risk is represented by the ICCDP and the ICLERP metrics referenced in RG 1.177.	Yes
Technical Specification Initiative 5b Risk-Informed Surveillance Frequencies	Nuclear Energy Institute report 04-10, Revision 1, quantitatively evaluates the change in total risk (including internal and external hazards contributions) in terms of CDF and LERF for both the individual risk impact of a proposed change in surveillance frequency and the cumulative impact from all individual changes to surveillance frequencies.	Yes
Maintenance Rule 10CFR 50.65 (a)(2)	Utilizes importance FV and RAW importance measures calculated from a PRA.	Yes
Maintenance Rule 10CFR 50.65 (a)(4)	Utilizes a risk monitor to evaluate the risk associated with a change in the plant configuration due to performing maintenance. Quantitative thresholds for risk management actions may be established by considering the magnitude of increase of the CDF (and/or large early release frequency) for the maintenance configuration. This is defined as the incremental CDF (ICDF), or incremental LERF (ILERF). The ICDF is the difference in the "configuration-specific" CDF and the baseline (or the zero maintenance) CDF. The configuration-specific CDF is the annualized risk rate with the out-of-service unavailability for SSCs set to one. The product of the ICDF (or ILERF) and duration is expressed as a probability (i.e., ICCDP and ILERP).	Yes

Table 3-1 Candidate risk-informed activities; licensee-initiated regulatory activities (Continuation).

Risk-informed activities	Type of risk metrics utilized/acceptance guidelines	Is PRA required?
Special Treatment 10CFR 50.69	Utilizes importance FV and RAW importance measures calculated from a PRA. Also utilizes the average annual CDF and LERF calculated from the baseline PRA and the change in CDF and LERF evaluated with proposed modifications to the plant design or operation reflected in the PRA.	Yes
ECCS Requirements 10CFR 50.46 (a)[10]	Utilizes pipe break frequency data to determine a transition break size.	No
Pressurized Thermal Shock 10CFR 50.61a	Utilizes the average annual CDF and LERF calculated from the baseline PRA. The total change in risk should be consistent with the acceptance guidelines of RG 1.174.	Yes
Generic Issue – 191	Utilizes the average annual CDF and LERF calculated from the baseline PRA. The total change in risk should be consistent with the acceptance guidelines of RG 1.174.	Yes
Generic Issue – 199	Utilizes the average annual CDF and LERF calculated from the baseline seismic PRA. The total change in risk should be consistent with the acceptance guidelines of RG 1.174.	Yes
Generic Issue - 204	Utilizes the average annual CDF and LERF calculated from the baseline external flood PRA. The total change in risk should be consistent with the acceptance guidelines of RG 1.174.	Yes

3.2   Step A-2: Application of PRA Results

The purpose of this step is to determine how the PRA results are used to support the decision under consideration. Addressing the uncertainties in the PRA is dependent on how the results are being used. Examples of different uses include:

- The risk metrics from the PRA are continuously being evaluated such that at any time, the risk associated with the current configuration is known (i.e., a risk monitor). Decisions that use a PRA in this way are not subject to the approach for the treatment of uncertainties discussed in this NUREG.

- The decision under consideration is based on reviewing the PRA risk results against specified regulatory acceptance guidelines. These guidelines may involve a change in the plant risk profile (e.g., ΔCDF) or the baseline risk metric (e.g., CDF). Decisions that use a PRA in this way are subject to the approach for the treatment of uncertainties

---

[10] This alternative regulation has not been approved by the Commission.

discussed in this NUREG. The exceptions are risk activities that only utilize importance measures.

- The risk significance of a decision is being evaluated at the time of occurrence of an event. For these time-sensitive situations, there is not enough time to evaluate the impact of uncertainties using the approach in this NUREG. As such, other means are used to address the uncertainties. Typically, licensee-initiated risk-informed activities do not fit into this category. A Notice of Enforcement Discretion is an example of an activity that may fit into this category.

Table 3-2 shows the list of activities from Table 3-1 and indicates how they utilize PRA results. Those license-initiated activities that are subject to the approach for addressing uncertainty discussed in this NUREG are identified in the last column of the table. For those activities that are not subject to the process in this NUREG, see Section 3.4. It is re-iterated that the list in Table 3-2 may not reflect evaluation of all possible risk-informed activities.

Table 3-2 Candidate risk-informed activities subject to the guidance in this NUREG.

Risk-informed activity	When are the risk results are used?			Subject to uncertainty assessment process in this NUREG?
	Continuously evaluated?	Evaluated to support an initiative?	Evaluated at the time of an event?	
Component Risk Ranking (Motor-Operated Valves)	No	Yes, but only uses importance measures	No	No
Mitigating Systems Performance Index (MSPI)	No	Yes, but only uses importance measures	No	No
NFPA 805 Fire Protection 10CFR 50.48(c)	No	Yes	No	Yes
Notice of Enforcement Discretion (NOED)	No	Possibly	Yes	No
Regulatory Guide 1.174	No	Yes	No	Yes
Regulatory Guide 1.175	No	Yes	No	Yes
Regulatory Guide 1.177	Yes	Yes	No	Yes, except for risk monitor
Regulatory Guide 1.178	No	Yes	No	Yes

Table 3-2 Candidate risk-informed activities subject to the guidance in this NUREG
(Continuation).

Risk-informed activity	When are the risk results are used?			Subject to uncertainty assessment process in this NUREG?
	Continuously evaluated?	Evaluated to support an initiative?	Evaluated at the time of an event?	
Significance Determination Process (SDP)	No	Possibly	No	No
Technical Specification Initiative 4b Risk-Informed Completion Times	Yes	Yes	No	Yes
Technical Specification Initiative 5b Risk-Informed Surveillance Frequencies	No	Yes	No	Yes
Maintenance Rule 10CFR 50.65 (a)(2)	No	Yes, but only uses importance measures	No	No
Maintenance Rule 10CFR 50.65 (a)(4)	Yes	No	No	No
Special Treatment 10CFR 50.69	No	Yes	No	Yes
Pressurized Thermal Shock 10CFR 50.61a	No	Yes	No	Yes
Generic Issue – 191	No	Yes	No	Yes
Generic Issue – 199	No	Yes	No	Yes
Generic Issue - 204	No	Yes	No	Yes

3.3   Summary of Stage A

This section provides guidance to both the licensee and the NRC staff on determining whether the approach for treating PRA uncertainties, as provided in this NUREG, should be used for the risk-informed activity (i.e., the decision) under consideration. The guidance involves determining the whether the results from a PRA are used in the application and how the results being used to support the decision.

Once it is determined that the risk-informed application requires a PRA and is thus subject to the guidance in this NUREG, the licensee proceeds to Stage B where the scope and level of detail of the PRA needed for a risk-informed decision is evaluated. However, as stated earlier, the process described in this NUREG is generic in nature and is generally applicable to all risk-informed decisions. Section 3.4 describes how this guidance can be applied generically.

3.4 Generic Application of Process

For risk-informed activities, the uncertainties associated with the risk assessment need to be addressed. The detailed manner in which they are addressed are dependent on the activity and the type of risk assessment. Nonetheless, the guidance provided on the uncertainty identification and characterization process and on the process of factoring the results into the decisionmaking is generic. The subsequent sections of this report provide the detailed guidance for specific risk-informed activities utilizing PRA insights for nuclear power plants. This section provides a summary of a more generic process that can involve different risk assessment methods and applications to different types of facilities.

The generic process is comprised of the following three steps:

- understanding the risk-informed activity
- understanding the sources of uncertainty
- addressing the uncertainties in the decisionmaking

3.4.1 Understanding the Risk-Informed Activity

It is necessary to understand the risk-informed decision under consideration; that is, understand what aspect of a plant design, operation, or performance is being assessed. Part of this understanding is determining whether a risk analysis is capable of providing the needed risk results to aid in the decision. This determination is three-fold: (1) the results from the risk assessment needed to make the risk-informed decision are identified, (2) an appropriate risk assessment method that can provide the necessary results is selected, and (3) the scope and level of detail of the risk assessment is adequate to evaluate the decision from a risk perspective.

When using the results of a risk assessment to support a risk-informed decision, the first step is to identify what results are needed to evaluate the specific aspect of plant design or operation being assessed, and how they are to be used to inform the decision. The results needed are generally formulated in terms of acceptance guidelines, which may be either numerical or qualitative in nature. Numerical acceptance guidelines can include specific metrics such as CDF or more general metrics such as the potential for fatalities. Qualitative acceptance guidelines can involve identification of important hazards and adequate prevention and mitigation measures.

The required risk information needed to compare against the acceptance guidelines can determine what type of risk assessment must be utilized. In most cases, numerical acceptance guidelines are utilized in risk-informed regulatory actions. For these types of actions, a quantitative risk assessment approach such as PRA is required. However, there can be occasions where qualitative risk insights are sufficient and thus a qualitative risk method (e.g., one of the methodologies used in Integrated Safety Assessments of fuel cycle facilities) could be sufficient to make a risk-informed decision.

The required risk results as well as the aspect of the plant design or operation being addressed in the risk-informed application determines the scope and level of detail needed in the risk assessment. The scope of the risk assessment can include contributions from different hazard groups (both internal and external to the facility) as well as during different modes of operation or facility configurations. In some cases, some hazard groups or plant operational states can be qualitatively shown to be clearly irrelevant to assessing the change in risk associated with the decision. The level of detail of the risk assessment can vary due to choices in the modeling assumptions and approximations made in order to limit the need for potentially resource intensive detailed analysis. Less detailed models can result conservative results which must be considered when using the results to make a risk-informed decision.

3.4.2 Understanding the Sources of Uncertainty

The impact of parameter, model, and completeness uncertainty on the risk assessment results must be identified in order to make proper risk-informed decisions. In general, identification and evaluation of these uncertainties for risk assessments other than PRA follows the same approach as provided in Sections 5, 6, and 7 of his report. Parameter uncertainty relates to the uncertainty in the input parameters in the risk assessment model and is only pertinent in quantitative risk assessment methodologies and can be evaluated by propagation of individual parameter uncertainty distributions through the risk assessment model. Completeness uncertainty arises from missing scope or level of detail items not included in the risk assessment model and can be evaluated using screening approaches. Model uncertainty is addressed in the following paragraph.

In order to identify the sources of model uncertainty and related assumptions that could have an impact on the risk results, the significant contributors to the risk results need to be identified. This can be accomplished by organizing the results needed from the risk assessment in such a way that they can be compared to the acceptance guidelines associated with the risk-informed decision. The significant contributors can be identified by decomposing the results in a hierarchical approach. For example, the risk results can be evaluated first for different hazard groups, followed by an identification of the significant accident sequences, and then down to the limit of the model (i.e., basic events).

The analysis of the significant contributors to a risk assessment results in an identification of the subset of the relevant sources of model uncertainty that could have an impact on the results. In the context of decision making, it is necessary to assess whether these uncertainties have the possibility of changing the evaluation of risk significantly enough to alter a decision. This can be done be performing sensitivity studies on significant model uncertainties and assumptions.

3.4.3 Addressing the Uncertainties in the Decisionmaking

When using results of a risk assessment to make a risk-informed decision, it is important to understand how uncertainties can impact the decision. For quantitative risk assessments, parameter uncertainty can be addressed in terms of a probability distribution on the numerical results of the risk assessment, and it is straightforward to compare a point value, be it the mean, the 95th percentile, or some other representative value, with a numerical acceptance guideline.

When the risk assessment is not complete, the missing scope or level of detail items must be addressed in the decision process. Screening assessments can be utilized to show that the missing items are not important to the decision. In some cases, the risk-informed activity can be altered such that the missing scope items do not affect the decision process.

With regard to model uncertainties, the results of sensitivity studies can confirm that the acceptance guidelines are still met even under the alternative assumptions (i.e., conclusion generally remains the same with respect to the acceptance guideline). The principle of risk-informed regulation dealing with acceptable risk impact can be determined to be met with confidence.

4.   STAGE B — ASSESSING PRA SCOPE AND LEVEL OF DETAIL

This section provides guidance to the licensee on determining the scope and level of detail of a probabilistic risk assessment (PRA) needed to support a risk-informed application. In Stage A, it is determined that the treatment of uncertainty for the risk-informed activity under consideration fits within the scope of the guidance provided in this NUREG. The goal of Stage B is to determine if the PRA has the scope and level of detail necessary to support the application. In this context, the term "PRA" is meant to refer to either an integrated model that includes all hazard groups and plant operating states (POS), or multiple PRA models that address the different hazard groups and POSs.

The required PRA scope and level of detail can vary for different risk-informed activities. It is important that the PRA address all important contributors to risk that can be affected by a proposed risk-informed activity. Similarly, the treatment of uncertainties can also vary and can be addressed per the guidance in this NUREG.

4.1   Overview of Stage B

At this stage, the analyst is determining whether the PRA scope and level of detail is sufficient to support the risk-informed decision under consideration. This is accomplished by the three steps, as illustrated in Figure 4-1 and described below.

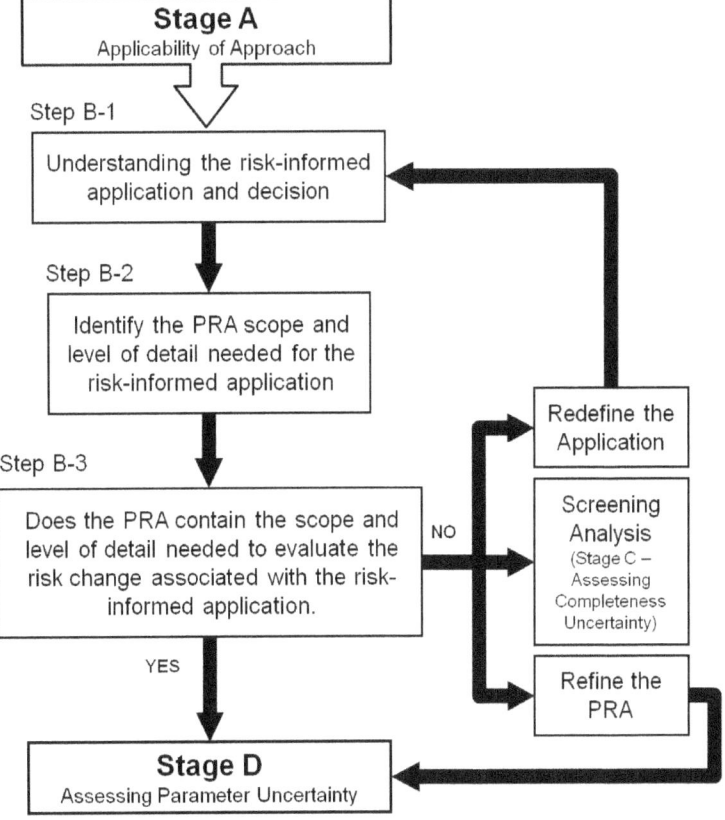

Figure 4-1 Overview of Stage B

- Step B-1: Understanding the risk-informed application and decision. It is essential to understand what structures, systems, and components (SSCs) and activities would be affected by a proposed application. In addition, a key aspect in this process is identifying the risk metrics needed to support the application.

- Step B-2: Identify the PRA scope and level of detail needed for the risk-informed application. By understanding the cause-and-effect relationship between the application and the SSCs and plant activities affected by the proposed risk-informed activity, the scope of the PRA required to evaluate the risk implications of the application can be determined. It is then necessary to identify if the existing PRA contains the elements that would be affected by an application and are important to evaluating the risk measure(s) applicable to the decision.

- Step B-3: Address the missing PRA scope or level of detail needed for the risk-informed application. If the PRA does not contain the scope and level of detail needed to evaluate the risk change associated with a risk-informed application, the licensee can choose one of the following options: (1) upgrade the PRA to include the required scope and level of detail, (2) redefine the application such that the missing PRA scope or level of detail is not needed in the evaluation process, or (3) perform a conservative/bounding assessment of the missing items to determine if they are significant to the decision.

These three steps are described in the following sections.

4.2   Step B-1: Understanding the Risk-Informed Application and Decision

The purpose of this step is to provide guidance to the licensee for determining the aspects of the plant design and operation that will be affected by a proposed risk-informed application. This is accomplished by identifying all aspects of the plant configuration (i.e., SSCs), operation, or requirements that may be affected by the proposed plant change. All impacts of a proposed application including the effect on the prevention and mitigation of transients or accidents are identified. This includes effects on both safety and non-safety related features.

When using the results of a risk assessment to support a risk-informed decision, the first step is to identify what results are needed and how they are to be used to inform the decision. The results needed are generally formulated in terms of acceptance guidelines (i.e., the results needed from the risk assessment are organized in such a way that they can be compared to the acceptance guidelines associated with the risk-informed decision). For a regulatory application, the acceptance guidelines can be found in the corresponding regulatory guide or in industry documents that are endorsed by the staff. Acceptance guidelines can vary from decision to decision, but most likely they will be stated in terms of the numerical value or values of some risk metric or metrics. The metrics commonly used include:

- core damage frequency (CDF)

- large early release frequency (LERF)

- change in CDF (ΔCDF) or change in LERF (ΔLERF)

- conditional core damage probability or conditional large early release probability

- incremental core damage probability or incremental large early release probability

- various importance measures such as Fussell-Vesely (FV), risk achievement worth (RAW), and Birnbaum

The acceptance guidelines also should include guidance on how the metric is to be calculated, in particular with regard to addressing uncertainty. In addition, when defining the metrics and the acceptance guidelines, it is necessary to define the scope of risk contributors that should be addressed.

4.3   Step B-2: Identify the PRA Scope and Level of Detail Needed to Support the Risk-Informed Decision

The required PRA scope and level of detail are dictated by the decision under consideration. This determination is generally accomplished by considering the cause-and-effect relationship between the application and its impact on the plant risk. A proposed application can impact multiple SSCs in various ways. Consequently, the application can require changes to one or more PRA technical elements. Depending on the application, these modifications could manifest as changes to parameters in the PRA model; introduction of new events; or changes in the logic structure. The following are additional examples of modifications to the PRA that might be needed as a result of an application:

- introduces a new initiating event

- requires modification of an initiating event group

- changes to a system success criterion

- requires addition of new accident sequences

- requires additional failure modes of SSCs

- alters system reliability or changes system dependencies

- requires modification of parameter probabilities

- introduces a new common cause failure mechanism

- eliminates, adds, or modifies a human action

- changes important results used in other applications, such as importance measures

- changes the potential for containment bypass or failure modes leading to a large early release

- changes the SSCs required to mitigate external hazards such as seismic events

- changes the reliability of systems used during low power and shutdown (LPSD) modes of operation

Once the effect of the application on the PRA model is determined, the PRA model is then reviewed to determine if it has the needed scope and level of detail defined by the application. Not all portions of a full-scope PRA will always be required to evaluate an application. Furthermore, some portions of the required PRA scope may not be important to the decision process.

For many risk-informed applications, application-specific guidance documents already exist that provide guidance on using a PRA model to address the issue (e.g., RG 1.174, RG 1.177 [NRC, 2002; NRC, 1998b], Nuclear Energy Institute (NEI) 00-04 [NEI, 2005b] and the Electric Power Research Institute (EPRI) PSA Applications Guide [EPRI, 1995]). For some applications, only a portion of the complete PRA model is needed. For other applications, such as the identification of the significant contributors to risk, the complete PRA is needed. In addition, the RG 1.174 acceptance guidelines are structured so that even if only a portion of the PRA results are required to assess the change in CDF and LERF (depending on the magnitude of that change), an assessment of the base PRA risk metrics (e.g., base CDF and LERF) may be needed.

Sections 4.3.1 and 4.3.2 provide guidance on selecting the required PRA scope and level of detail, respectively. In addition, the importance of the variability in the level of detail and approximation utilized in the evaluation of different hazard groups and POSs is important in making a decision. Section 4.3.3 discusses the influence of the level of detail of different portions of a PRA on the aggregation of results used to make a risk-informed decision.

4.3.1 PRA Scope

The scope of the PRA is defined in terms of the following:

- the metrics used to evaluate risk

- the POSs for which the risk is to be evaluated

- the types of hazard groups and initiating events that can potentially challenge and disrupt the normal operation of the plant and, if not prevented or mitigated, would eventually result in core damage, a release, and/or health effects

For regulatory applications, the scope of risk contributors that needs to be addressed includes all hazard groups and all plant operational states that are relevant to the decision. For example, if the decision involves only the at-power operational state, then the low-power and shutdown operational states need not be addressed.

Risk metrics are the end-states (or measures of consequence) quantified in a PRA to evaluate risk. In a PRA, different risk metrics are generated by Level 1, limited Level 2, Level 2, or Level 3 PRA analyses.

- Level 1 PRA: Involves the evaluation and quantification of the frequency of the sequences leading to core damage. The metric evaluated is CDF.

- Limited Level 2 PRA: Involves the evaluation and quantification of the mechanisms and probabilities of subsequent radioactive material releases leading to large early releases from containment. The metric evaluated is the large early release frequency.

- Level 2 PRA: Involves the evaluation and quantification of the mechanisms, amounts, and probabilities of all the subsequent radioactive material releases from the containment. The metrics evaluated include the frequencies of different classes of releases, which include large release, early release, large late release, small release.

- Level 3 PRA: Involves the evaluation and quantification of the resulting consequences to both the public and the environment from the radioactive material releases. The metrics are typically measures of public risk that include frequencies of early fatalities and latent cancer fatalities.

The risk metrics relevant to the decision are defined by the acceptance guidelines associated with the decision. For example, for a licensing-basis change, RG 1.174 defines the risk metrics as CDF, LERF, ΔCDF, and ΔLERF. Therefore, if the acceptance guidelines use CDF, LERF, ΔCDF, and ΔLERF, then the PRA scope generally should address these risk parameters. However, circumstances may arise where the ΔCDF associated with an application is small enough that it also would meet the ΔLERF guidelines. In this case, explicit modeling of the LERF impacts of the application may not be necessary.

Plant operating states (POSs) are used to subdivide the plant operating cycle into unique states such that the plant response can be assumed to be the same for all subsequent accident-initiating events. Operational characteristics (such as reactor power level; in-vessel temperature, pressure, and coolant level; equipment operability; and changes in decay heat load or plant conditions that lead to different success criteria) are examined to identify those relevant to defining POSs. These characteristics are used to define the states, and the fraction of time spent in each state is estimated using plant-specific information. The risk perspective is based on the total risk associated with the operation of the nuclear power plant, which includes not only full-power operation but also other operating states such as low-power and shutdown conditions.

The impact of the application determines the POSs to be considered. The various operating states include at power, low power, and shutdown. Not every application necessarily will impact every operating state. In deciding this aspect of the required scope, the SSCs affected by the application are identified. It is then determined if the affected SSCs are required to prevent or mitigate accidents in the different POSs and if the impact of the proposed plant change would impact the prevention and mitigation capability of the SSCs in those POSs. A plant change could affect the potential for an accident initiator in one or more POSs or reduce the reliability of a component or system that is unique to a POS or required in multiple POSs. Once the cause-and-effect relationship of the application on POSs is identified, the PRA model is reviewed to determine if it has the scope needed to reflect the effect of the application on plant risk.

Initiating events perturb the steady state operation of the plant by challenging plant control and safety systems whose failure could potentially lead to core damage and/or radioactivity release. These events include failure of equipment from either internal plant causes (such as hardware faults, operator actions, floods, or fires) or external plant causes (such as earthquakes or high winds). The challenges to the plant are classified into **hazard groups**, which are defined as a group of similar hazards that are assessed in a PRA using a common approach, methods, and likelihood data for characterizing the effect on the plant. Typical hazard groups for a nuclear power plant PRA include internal events, seismic events, internal fires, internal floods, and high winds.

The impact of the application determines the types of hazard groups to be considered. In deciding this aspect of the required scope, the process is similar to that described above for POSs. The SSCs affected by the application are determined, and the resulting cause-and-effect relationships are identified. It is then determined if the affected SSCs are required to prevent or mitigate both internal and external hazard groups and if the proposed plant change would affect that capability. The impact of the proposed plant change on the SSCs could introduce new accident initiating events caused by a hazard, affect the frequency of initiators, or affect the reliability of mitigating systems required to respond to multiple initiators. Once the cause-and-effect relationship on the accident-initiating events is identified, the PRA model is reviewed to determine if it has the scope needed to reflect the effect of the application on plant risk.

For example, consider an application that involves a licensing-basis change where a seismically qualified component is being replaced with a non-qualified component. If the new component's reliability is not changed relative to its response to non-seismic events, the non-seismic part of the PRA is not impacted and only the seismic risk need be considered; that is, the other contributors to the risk (e.g., fire) are not needed for this application. If, on the other hand, the reliability of the new component is changed relative to its response to non-seismic events, the non-seismic part of the PRA may be impacted and, therefore, needs to be included in the scope.

As another example, if an application does not affect the decay heat systems at a plant, an evaluation of loss-of-heat removal events during low-power shutdown would not be required. However, the assessment of other events, such as drain-down events, may still be required.

4.3.2 Level of Detail

The level of detail of a PRA is defined in terms of (1) the degree to which the potential spectrum of scenarios is discretized and (2) the degree to which the actual plant is modeled. A number of decisions made by the analyst determine the level of detail included in a PRA. These decisions include, for example, the structure of the event trees, the mitigating systems that should be included as providing potential success for critical safety functions, the structure of the fault trees, and the screening criteria used to determine which failure modes for which SSCs are to be included. The degree of detail required of the PRA for an application is determined by how it is intended to be used in the application. Although it is desirable for a PRA is to be as realistic as practicable, it is often necessary to find a compromise between realism and practicality, as discussed below.

The logic models of a PRA (i.e., the event trees and fault trees) are a simplified representation of the complete range of potential accident sequences. For example, modeling all the possible initiating events or all the ways a component could fail would create an unmanageably complex and unwieldy model. Consequently, simplifications are achieved by making approximations. As an example, initiating events are consolidated into groups whose characteristics bound the characteristics of the individual members. As another example, when developing an accident sequence timeline, a representative sequence is generally chosen that assumes that all the failures of the mitigating systems occur at specific times (typically the time at which the system is demanded). However, in reality, the failures could occur over an extended time period (e.g., the system could fail at the time demanded or could fail at some later time). Developing a model that represents all the possible times the system could fail and the associated scenarios is not practical. The time line is used, among other purposes, to provide input to the human reliability analysis. Typically, a time is chosen that provides the minimum time for the operator to receive the cues and to complete the required action. This minimized time maximizes the probability of failure. This simplification, therefore, leads to an uncertainty in the evaluation of risk that is essentially unquantifiable without developing more detailed models that model more explicitly the different timelines. The basis for the compromise is the assumption that the simplification provides adequate detail for the intended purpose of the model. It also is generally assumed that the simplification results in a somewhat conservative assessment of risk.

The degree to which plant performance is represented in the PRA model also has an effect on the precision of the evaluation of risk. For each technical element of a PRA, the level of detail may vary by the extent to which the following occur:

- plant systems and operator actions included in modeling the plant design and operation

- plant-specific experience and the operating history of the plant's SSCs are incorporated into the model

- realism is incorporated in the deterministic analyses to predict the expected plant responses

The level of detail in the way the logic models are discretized and the extent to which plant representation is modeled is at the discretion of the PRA analyst. The analyst may screen out initiating events, component failure modes, and human failure events so that the model does not become encumbered with insignificant detail. For example, not all potential success paths may be modeled. However, a certain level of detail is implicit in the requirements of the ASME/ANS PRA standard. Although an analyst chooses the level of detail, the PRA model needs to be developed enough to correctly model the major dependencies (e.g., those between front line and support systems) and to include the significant contributors to risk. Nonetheless, the coarser the level of detail, the less precise is the estimate, resulting in uncertainty about the predictions of the model. The generally conservative bias that results could be removed by developing a more detailed model.

In many cases, the level of detail will be driven by the requirements of the application for which the PRA is being used. In particular, the PRA model needs to adequately reflect the cause-effect relationship associated with an application. As an example, PRAs typically do not include failures of passive components such as pipes in the system fault tree models. For an inservice inspection application under Regulatory Guide 1.178, risk information is used to categorize piping segments into either high- or low-safety significant classifications. Although one could add piping failures in the PRA model, other components in each piping segment that are included in the PRA model can serve as surrogates for determining the piping safety significance. An important point from this example is that the PRA must have the necessary level of detail to evaluate the impact of the application.

The technical requirements of the ASME and American Nuclear Society (ANS) PRA standard [ASME/ANS, 2009] provide a means to establish that an acceptable level of detail exists for a base PRA, independent of an application. The use of screening analyses (either qualitative or quantitative) is an accepted technique in PRA for determining the level of detail included in the analysis. Section 5.2 includes a number of examples of screening analyses. However, it is recognized that the detail included in the PRA model may not be needed for a given application, although the minimum level of detail may not be sufficient in other cases. The level of detail needed is that detail required to assess the effect of the application (i.e., the PRA model needs to be of sufficient detail to ensure the impact of the application can be assessed). Again, the impact of the application is achieved by reviewing the cause-and-effect relationship between the application and its impact on the PRA model.

4.3.3 Aggregation of Results from Different PRA Models

For all applications, it is necessary to consider the contributions from the applicable hazard groups and/or plant operational states in quantifying the risk metrics such as CDF, LERF, or an importance measure. Because the hazard groups and plant operating states are independent, addition of the contributions is mathematically correct. However, several issues should be considered when establishing the required PRA scope and level of detail, and in combining the results from different hazard groups and POSs.

When combining the results of PRA models for several hazard groups (e.g., internal events, internal fires, seismic events) as required by many acceptance guidelines, the level of detail and level of approximation may differ from one hazard group to the next with some being more conservative than others. This modeling difference is true even for an internal events, at-power PRA. For example, the evaluation of room cooling and equipment failure thresholds can be conservatively evaluated leading to a conservative time estimate for providing a means for alternate room cooling. Moreover, at-power PRAs follow the same general process as used in the analysis of other hazard groups, with regard to screening: low-risk sequences can be modeled to a level of detail sufficient to prove they are not important to the results.

Significantly higher levels of conservative bias can exist in PRAs for external hazards, LPSD, and internal fire PRAs. These biases result from several factors, including the unique methods or processes and the inputs used in these PRAs as well as the scope of the modeling. For example, the fire modeling performed in a fire PRA can use simple scoping models or more sophisticated computer models or a mixture of methods and may not mechanistically account for all factors such as the application of suppression agents. Moreover, in an effort to reduce the number of cables that have to be located, fire PRAs do not always credit all mitigating systems. To a certain level, conservative bias will be reduced by the development of detailed models and corresponding guidance for the analysis of external hazards, fires, and LPSD that will provide a similar level of rigor to the one currently used in internal events at-power PRAs. However, as with internal events at-power PRAs, the evaluation of some aspects of these other contributors will likely include some level of conservatism that may influence a risk-informed decision.

The level of detail, scope, and resulting conservative biases in a PRA introduces uncertainties in the PRA results. Because conservative bias can be larger for external hazards, internal fire, and LPSD risk contributors, the associated uncertainties can be larger. However, a higher level of uncertainty does not preclude the aggregation of results from different risk contributors; but it does require that sources of conservatism having a significant impact on the risk-informed application be recognized.

The process of aggregation can be influenced by the type of risk-informed application. For example, it is always possible to add the CDF or LERF, or the changes in CDF or changes in LERF contributions from different hazard groups for comparison against corresponding acceptance guidelines. However, in doing so, one should always consider the influence of known conservatism when comparing the results against the acceptance guidelines, particularly if they mask the real risk contributors (i.e., distort the risk profile) or result in exceeding the guidelines. If the acceptance guidelines are exceeded due to a conservative analysis, then it may be possible to perform a more detailed, realistic analysis to reduce the conservatism and uncertainty. For applications that use risk importance measures to categorize or rank SSCs according to their risk significance (e.g., revision of special treatment), a conservative treatment of one or more of the hazard groups can bias the final risk ranking. Moreover, the importance measures derived independently from the analyses for different hazard groups cannot be simply added together and thus would require a true integration of the different risk models to evaluate them.

Sections 8 and 9 discuss how the NRC staff evaluates the acceptability of the licensee's aggregation of the results from different risk contributors. To facilitate this effort, it is best that results and insights from all of the different risk contributors relevant to the application be provided to the decisionmaker in addition to the aggregated results. This information will allow for consideration of at least the main conservatisms associated with any of the risk contributors and will help focus the decisionmaker on those aspects of the analysis that have the potential to influence the outcome of the decision.

4.4 Step B-3: Address the Missing PRA Scope or Level of Detail Needed for the Risk-Informed Application.

The outcome from Step B-2 is the identification of the PRA scope and level of detail necessary to evaluate the risk associated with a risk-informed application. Once this determination is made, the existing PRA is evaluated to determine if it is sufficient for the application. If the application is sufficient, the PRA or portion of the PRA is modified as necessary to evaluate the application.

However, if the PRA scope or level of detail is insufficient to evaluate the proposed application, the licensee will have to take some action. Three possible actions are: (1) upgrade the PRA to include the required scope and level of detail, (2) redefine the application such that the missing PRA scope or level of detail is not needed in the evaluation process, or (3) perform a conservative/bounding assessment of the missing PRA scope or level of detail to determine if they are significant to the decision. Although these options are discussed in Stage F as part of the licensee's application development process, they are briefly described below.

Upgrading the PRA to evaluate a missing scope item may be required if it is significant to the decision. The effort required to modify the level of detail in the PRA will generally be less than to evaluate a missing hazard group or POS. If a hazard group or POS is determined to be significant to a risk-informed application, the Commission has directed that the PRA used to support that application be performed against an available, staff-endorsed PRA standard for that specific hazard group [NRC, 2003e].

As an alternative, the risk-informed application can be redefined such that the scope of a risk-informed change is restricted to those areas supported by the existing PRA's scope and level of detail. For example, if the PRA model does not address shutdown modes of operation, the change to the plant could be limited such that any SSCs that would be expected to be used to mitigate the risk from accidents during shutdown would be unaffected by the proposed plant change. In this way, the contribution to risk from events occurring during shutdown would be unchanged by the application and there would be no need to evaluate the risk during shutdown.

For a given application, the PRA does not need to be upgraded if it can be demonstrated that the missing scope or missing level of detail does not impact the risk insights supporting the risk-informed application. This can be demonstrated by performing a bounding, conservative, or a realistic but limited screening assessment of the missing scope or level-of-detail items. Section 5.2 provides guidance on performing these types of screening assessments.

4.5 Summary of Stage B

This section provides guidance to the licensee for determining the scope and level of detail of a PRA needed to support a risk-informed application. Stage B is entered when it has been determined, in Stage A, that the treatment of uncertainty for the risk-informed activity under consideration fits within the scope of the guidance provided in this NUREG.

The required PRA scope and level of detail can vary for different risk-informed activities. It is important that the PRA address all important contributors to risk that can be affected by a proposed risk-informed activity. Similarly, the treatment of uncertainties can also vary and can be addressed per the guidance in this NUREG.

Stage B provides guidance to the licensee on determining whether the PRA scope and level of detail is sufficient to support the risk-informed decision under consideration. This is accomplished by the understanding what SSCs and activities would be affected by a proposed application and identifying the risk metrics needed to support the application. By understanding the cause-and-effect relationship between the application and the SSCs and plant activities affected by the proposed risk-informed activity, the scope of the PRA required to evaluate the risk implications of the application can be determined.

If the PRA does not contain the scope and level of detail needed to evaluate the risk change associated with a risk-informed application, the licensee will have to address this fact in Stage F as part of the application development process. Two possible options the licensee can choose is to either upgrade the PRA to include the required scope and level of detail or redefine the application such that the missing PRA scope or level of detail is not needed in the evaluation process. If the licensee chooses either of these two options, the resulting PRA is evaluated per the guidance in Stage D. Alternatively, the licensee can choose to evaluate the significance of missing scope and level-of-detail items to the decision under consideration by performing a conservative/bounding assessment per the guidance in Stage C.

5. STAGE C — ASSESSING COMPLETENESS UNCERTAINTY

This section provides guidance to the licensee on how to address the completeness of the probabilistic risk assessment (PRA) results that are used in support of risk-informed applications. In Stage B, the adequacy of the scope and level of detail of the PRA is determined relative to the decision under consideration. The goal of Stage C is to describe how to address the scope and level-of-detail items that are not modeled in the PRA and, ultimately, determine whether those missing scope and level-of-detail items (e.g., a hazard group, an initiating event, a component failure mode, etc.) are significant to the decision under consideration.

As discussed in Section 4, a PRA used in a risk-informed application should be of sufficient scope and level of detail to support the risk-informed decision under consideration. Moreover, the risk from each significant hazard group should be addressed using a PRA model that is developed in accordance with an NRC-endorsed consensus standard for that hazard group. A significant hazard group (e.g., risk contributor) is one whose consideration can affect the decision, and therefore, needs to be factored into the decisionmaking process. However, some contributors can be shown to be insignificant or irrelevant, and therefore, can be screened from further consideration.

The process of addressing completeness uncertainty corresponds to Stage C of the overall process for the treatment of uncertainties.

5.1 Overview of Stage C

In Stage B, the scope and level of detail of the PRA model needed to support the application has been defined. At this stage, as noted above, the goal is to determine whether the missing scope or level-of-detail items are significant relative to the acceptance guidelines. This process involves two steps as illustrated in Figure 5-1 below.

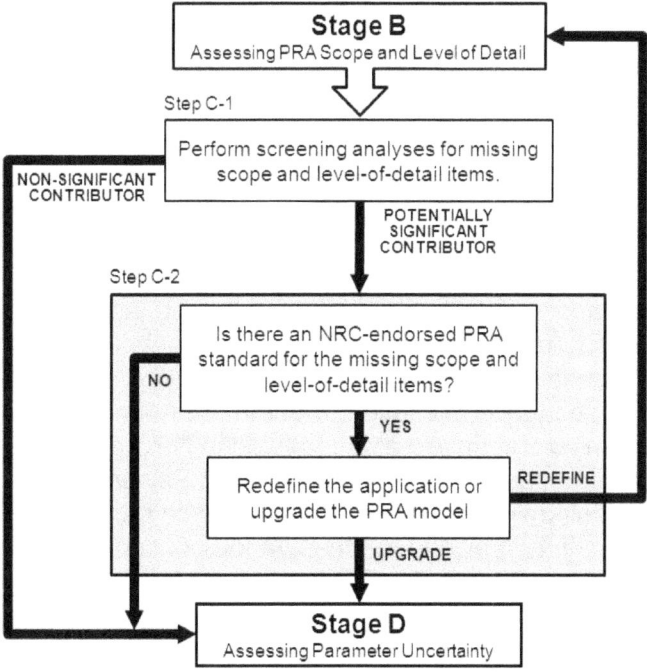

Figure 5-1 Overview of Stage C

- Step C-1: Perform screening analyses to determine the significance of the missing PRA scope or level of detail to the risk-informed decision.

- Step C-2: Determine if the PRA model needs to be updated or if the application needs to be modified to address the missing PRA scope or level of detail significant to the decision.

5.2 Step C-1: Perform Screening Analyses to Assess Significance

The purpose of this step is to provide guidance for determining whether the missing scope or level of detail of the PRA (as determined in Stage B) is risk significant to the decision under consideration. This determination is necessary because the licensee has initially decided not to upgrade the PRA or redefine the application in order to address the missing PRA scope or level-of-detail items.

The process of determining the risk significance of a missing scope or level of detail PRA item consists of performing a screening analysis that is used to demonstrate that the particular non-modeled PRA scope or level-of-detail item can be eliminated from further consideration in the risk-informed decision. Screening analyses are either qualitative or quantitative in nature or a combination of both types.

A qualitative screening analysis demonstrates that the scope or level-of-detail item cannot impact plant risk or is not important to the application specific acceptance guidelines (e.g., change in core damage frequency (CDF) or large early release frequency (LERF) as a result of a proposed plant modification). For example, specific hazard groups will be eliminated for many plant sites based on the fact that it is physically impossible for the hazards to occur (e.g., an avalanche in the middle of the desert). An example of a scope item that is not important to the change in risk is the following: an application to change an at-power technical specification would not impact low-power/shutdown (LPSD) risk, and thus a lack of a LPSD PRA would not be an issue regarding the necessary PRA scope to address the application.

A quantitative screening analysis produces a conservative estimate of the risk or change in risk from the proposed plant modification related to a scope item. Examples of this include analyses that demonstrate that an initiating event has a very low frequency of occurrence or that a plant change does not significantly affect the unavailability of a system.

Whether a qualitative or quantitative screening approach is used, the screening process should use screening criteria that is appropriate for the item being screened. Moreover, in developing a PRA model, screening is an inherent part of the process. As such, the ASME/ANS PRA standard provides various screening criteria that were developed to ensure that risk-significant items (e.g., hazards, events, sequences, or failure modes) are not eliminated (see Table 5-1). In addition, acceptable screening criteria are also provided in NRC guidance documents. An example is the screening criteria specified in the joint Electric Power Research Institute (EPRI)/NRC fire PRA methodology [EPRI, 2005]. The guidance provided below is discussed in the context of the screening criteria provided in these various documents[11].

---

[11] The guidance in this Section on screening is consistent with the ASME/ANS PRA standard to be endorsed in a staff Interim Staff Guidance document to be published.

Table 5-1 Supporting requirements in the ASME/ANS PRA standard that address screening.

Part of standard	Screened Parameters						
	Hazards	Initiating Events	Component /Failure Modes	Human Errors	Failure Rates/ Fragilities	Accident Sequences	Plant Areas
2		IE-C6	SY-A15 SY-B1 SY-B13	HR-B1 HR-B2	DA-C3 DA-D5 DA-D6	QU-B2 QU-B3	
3		IFEV-A8				IFQU-A3	IFSO-A3 IFSN-A12 IFSN-A13
4			ES-A3 ES-A4				FSS-G2 QLS-A1 QLS-A2 QLS-A3 QLS-A4 QNS-A1 QNS-C1
5	SHA-I1				SFR-B1 SFR-E3 SFR-F3 SPR-B3 SPR-B4a		
6	EXT-B3 EXT-C1 EXT-D1 EXT-D2						
7		WPR-A6	WPR-A6 WPR-A7 WPR-A8	WPR-A6 WPR-A8	WPR-A6 WPR-A7	WPR-A6	
8		XFPR-A6	XFPR-A6 XFPR-A7 XFPR-A8	XFPR-A6 XFPR-A8	XFPR-A6 XFPR-A7	XFPR-A6	
9		XPR-A6	XPR-A6 XPR-A7 XPR-A8	XPR-A6 XPR-A8	XPR-A6 XPR-A8	XPR-A6	

For Step C-1, the screening and significance assessment process consists of the following steps:

- <u>Substep C-1.1</u>: Perform qualitative screening
- <u>Substep C-1.2</u>: Perform quantitative screening
- <u>Substep C-1.3</u>: Determine significance of unscreened scope items

The general process for screening missing PRA scope or level-of-detail items is a progressive process that can involve different levels and various combinations of qualitative and quantitative screening. In general, qualitative screening is performed prior to any quantitative screening analysis. When a missing PRA scope and level-of-detail items cannot be eliminated from further consideration in the decisionmaking process by a qualitative or quantitative screening analysis, the significance of that unscreened item to the risk-informed application must be determined.

5.2.1 <u>Substep C-1.1</u>: Qualitative Screening

Qualitative screening is used during the development of a PRA model. When employing this type of screening analysis, the analyst should use approved screening criteria to eliminate potential hazard or risk contributors (i.e., scope or level-of-detail items) from the PRA. The NRC staff is developing a position on acceptable screening criteria which will be published in an Interim Staff Guidance document. A principle being followed in developing the screening criteria is that the qualitative and quantitative screening criteria should be consistent since their purpose is to eliminate events that are negligible contributors to risk. Examples of acceptable qualitative screening criteria include the following:

- The contributor or hazard does not result in a plant trip (manual or automatic) or require an immediate plant shutdown, and does not impact any SSCs that are required for accident mitigation. This criterion can also be applied to the process of screening equipment compartments as part of internal fire and flood analyses.

- The hazard is of equal or lesser damage potential than the hazards for which the plant has been designed. This requires an evaluation of plant design bases in order to estimate the resistance of plant structures and systems to a particular hazard.

- The contributor or hazard cannot occur close enough to the plant to affect it. Application of this criterion must take into account the range of magnitudes of the hazard for the recurrence frequencies of interest.

- The contributor or hazard is included in the definition (i.e., evaluation) of another hazard or event.

- The contributor's or hazard's effect is slow to develop and it has been demonstrated that there is sufficient time to eliminate the source of the threat or to provide an adequate response so that there is no impact on the operational risk. For example, in the case of an initiating event, a random initiator such as a loss of heating, ventilation, and air conditioning can be screened if the event does not require the plant to transition to shutdown conditions until some specified length of time has elapsed. During this time, it can be shown with a high degree of reliability that the initiating event conditions can be detected and corrected before normal plant operation is required to be terminated (either automatically or administratively in response to a limiting condition of operation). Deterministic analyses are necessary to support the use of this criterion.

- The missing hazard or event can be eliminated if there are procedures to sufficiently recover from the event, as is the case with pre-accident human errors to restore equipment following test and maintenance. In this case, factors such as automatic realignment of equipment on-demand; the performance of post maintenance tests or position indicators in the control room that would reveal misalignment; and frequent equipment status checks can all be used as justification for screening out pre-accident human errors. However, there is an implied probabilistic (quantitative) argument associated with such screening (i.e., the probability of a pre-accident human error going unnoticed is reduced sufficiently if these screening factors are applicable).

Variations of these criteria may exist and, in some cases, require a supporting deterministic analysis. For example, screening of flood areas is allowed if no plant trip or shutdown would occur and if the flood area has flood mitigation equipment, such as drains, capable of preventing flood levels that would result in damage to equipment or structures. A deterministic analysis would be necessary in this case to show that the drains are of sufficient size to prevent the water level from all flood sources from rising to a level that would result in equipment or structure damage.

Additional qualitative screening criteria may be identified as applicable for specific applications. The bases for any criteria used to qualitatively eliminate missing scope and level-of-detail items from a PRA must be documented.

5.2.2 <u>Substep C-1.2</u>: Quantitative Screening

Quantitative screening is also used during the development of a PRA model. It is used, however, somewhat differently than was discussed for qualitative screening. Quantitative screening is used:

- to demonstrate that the risk from the missing scope or level-of-detail item is not an important contributor to risk

- in lieu of a detailed PRA model to estimate the risk contribution from the item

If a PRA scope item is not screened based on a quantitative analysis, the results of that analysis can be used in the application to provide a conservative risk estimate if the item is not significant to the decision, as determined in Step C-2. Different levels of quantitative analysis are used for screening or evaluating the importance of missing PRA scope and level of detail. The following list of the different types of quantitative screening analyses is organized from the highest to the lowest level of conservatism:

- bounding quantitative analysis
- conservative, but not bounding analysis
- realistic, but limited quantitative analysis

The screening process can progress from a bounding assessment to a realistic analysis with the intent of screening a scope or level-of-detail item from the PRA model. In any case, the process of quantitative screening must minimize the likelihood of omitting any significant risk contributors. This is accomplished by using suitably low quantitative screening criteria (i.e., criteria representing a small fraction of the frequency or probability of expected events).

Bounding Quantitative Analysis

In the context of a specific PRA scope or level-of-detail item, a bounding analysis provides an upper limit of the risk metrics and includes the worst credible outcome of all known possible outcomes that result from the risk assessment of that item. The worst credible outcome is the one that has the greatest impact on the defined risk metric(s). Thus, a bounding probabilistic analysis must be bounding both in terms of the potential outcome and the likelihood of that outcome. This definition is consistent with, but more inclusive than, that provided in the ASME/ANS PRA standard, which defines a bounding analysis as an, "Analysis that uses assumptions such that the assessed outcome will meet or exceed the maximum severity of all credible outcomes."

Performance of a bounding analysis utilizes available knowledge to set an upper limit on where the answer may realistically lie. When compared to a best estimate probabilistic evaluation, a bounding value may represent a 95%, 98%, or some other percentile of the best estimate value. However, it is not practical to establish a specified percentile in the definition of a bounding analysis since one could only meet that definition by performing a best estimate analysis. Instead, a bounding analysis can only provide a point estimate of the risk metric associated with a missing scope or level-of-detail item. To accomplish this, a bounding analysis can be broken down into individual constituent parts with bounding values, assumptions, and models utilized in each piece of the evaluation. For example, a bounding scenario may be defined utilizing a bounding initiator frequency, assumed failure of available mitigating systems, and a maximum possible release of hazardous material. If the uncertainty distribution is available for one of the parameters such as the initiator frequency, a value representing a high percentile (e.g., 95th percentile) could be selected as a bounding value.

Some examples of bounding analyses that affect the PRA level of detail, are assuming that all fires or floods in a specific area (maximum frequency) fails all equipment in that location (maximum consequences) combined with taking no credit for mitigation systems (e.g., fire suppression or floor drains). In the case of a reactor pressure vessel rupture, the assumption that rupture occurs below the core simplifies the modeling of that scenario in a PRA, since emergency core cooling systems cannot reflood the vessel. This assumption maximizes the associated CDF estimate.

How the above definition is applied is dependent on whether a bounding analysis is intended to bound the risk or screen the PRA item as a potential contributor to risk. If a bounding analysis is being used to bound the risk (i.e., determine the magnitude of the risk impact from an event), then both its frequency and outcome must be considered. Conversely, if a bounding analysis is being used to screen the event (i.e., demonstrate that the risk from the event does not contribute to the defined risk metric(s)), then the event can be screened based on frequency, outcome, or both, depending on the specific event. For example, an explosion hazard could be screened based on the fact that bounding deterministic analyses of possible explosions indicates that these explosions could not damage plant equipment. If an explosion hazard cannot be deterministically screened, then a bounding risk assessment of the possible explosion hazard would also require bounding its frequency.

Conservative, but not Bounding Analysis

A conservative analysis provides a result that may not be the worst result of a set of outcomes, but produces a quantified estimate of a risk metric that is greater than the risk metric estimate that would be obtained by using a best-estimate evaluation. There are different levels of conservatism employed by an analyst to address specific scope items where the highest level of conservatism is achieved by using a bounding analysis. For example, conservative human error probabilities (e.g., 0.3) may be used in a risk evaluation, whereas a bounding risk estimate would require setting the human error probabilities equal to 1.

The level of conservatism is characterized by the selection of the models and data, assumptions, as well as the level of detail, used to analyze the scope item. For example, the frequency of a seismic hazard could be evaluated using different levels of conservative methods, data, and assumptions. Any approximations or simplifications used in the screening process must result in conservative or bounding risk estimates. For example, all seismic events could conservatively be assumed to result in loss of offsite power (LOSP) transients or LOSP events combined with a small loss-of-coolant accident (LOCA), the latter assumption resulting in a more conservative model of seismic events, especially those of low magnitude.

For screening purposes, the level of conservatism used is generally the minimum required to generate a frequency, consequence, or risk estimate that is below established criteria (i.e., the level of conservatism may have to be reduced in order to screen out an item). When a less-than-bounding but conservative analysis does not result in the screening of an item, it may be necessary to perform a more detailed analysis to either screen the scope item or provide a more realistic estimate for use in the risk-informed application. It is possible that a specific PRA item could be screened using a combination of conservative and best-estimate models and data.

Quantitative screening can involve the use of conservative, deterministic analyses. An example of a deterministic analysis supporting a quantitative screening process is provided in the quantitative fire compartment screening process documented in NUREG/CR-6850 [EPRI, 2005]. This methodology describes using conservative fire modeling analyses to identify fire sources that can potentially cause damage to important equipment. This analysis allows for eliminating fire sources that cannot cause damage, thus reducing the compartment fire frequency that is used in the quantitative screening process.

Realistic, but Limited Quantitative Screening Analysis

A realistic, but limited, quantitative screening analysis is an iterative process of successive screening that uses the PRA technical elements as a guide. A progressive screening process may begin with screening out an initiating event entirely given its frequency is sufficiently low. However, if the frequency is not low enough, then specific sequences associated with the initiating event may be screened. This progressive screening process either uses the same criteria (e.g., frequency) throughout or uses different criteria applicable to specific accident sequences (e.g., frequency plus number of available systems). The progressive screening approach may end with a detailed best-estimate analysis of the PRA item in question, which may or may not result in screening.

Regardless of the type of quantitative screening used, the appropriate quantitative screening criteria must be used in the analysis. The screening criteria are either purely quantitative or incorporate both quantitative and qualitative components. They should include consideration of the individual risk contribution of a screened item and the cumulative risk contribution of all screened items. Special consideration should be given to those items that can result in containment bypass. Some of the criteria also use comparative information as a basis for screening (e.g., an item can be screened whose risk contribution is significantly less than the contribution from another item that results in the same impacts to the plant). Some examples of these criteria are provided below to illustrate the general nature of the screening criteria that can be utilized. The bases for any quantitative criteria used to screen out any scope items from a PRA for a specific risk-informed application must be documented.

- An initiating event can be screened if its frequency is less than 10^{-7}/yr and provided the event is not a high consequence event such as an interfacing system LOCA, containment bypass, or reactor vessel rupture. Alternatively, an initiating event can be screened if its frequency is less than 10^{-6}/yr and core damage could not occur unless two trains of mitigating systems are failed, independent of the initiator [ASME/ANS, 2009].

- A component may be excluded from a system model if the total failure probability of all the component failure modes, which result in the same effect on system operation, is at least two orders of magnitude lower than the highest failure probability of other components in the same system, whose failure modes also result in the same effect on system operation. A component failure mode can be excluded from the system model if its contribution to the total failure probability is less than 1 percent of the total failure probability for the component. This exclusion is permissible when the effect on system operation of the excluded failure mode does not differ from the effects of the included failure modes. However, if a component is shared among different systems (e.g., a common suction pipe feeding two separate systems), then these screening criteria do not apply [ASME/ANS, 2009].

- An internal flood event can be screened if it affects only components in a single system and if it can be shown that the product of the flood frequency and the probability of SSC failures, given the flood, is two orders of magnitude lower than the product of the following two parameters: (1) the non-flooding frequency for the corresponding initiating events in the PRA and (2) the random (non-flood induced) failure probability of the same SSCs that are assumed failed by the flood [ASME/ANS, 2009].

- A flood area can be screened if the product of (a) the sum of the frequencies of the flood scenarios for that area and (b) the bounding conditional core damage probability is less than 10^{-9}/yr [ASME/ANS, 2009].

- A fire compartment can be screened if the CDF is less than 10^{-7}/yr and LERF is less than 10^{-8}/yr [EPRI, 2005]. In addition, the cumulative risk estimate (either realistic or conservatively determined) for the screened fire compartments should be less than 10 percent of the total internal fire risk [NRC, 2009].

There is limited guidance available for performing bounding or conservative screening analyses (e.g., guidance for performing conservative assessments of external hazards is provided in NUREG/CR-4832 [SNL, 1992a] and NUREG/CR-4839 [SNL, 1992b]). Furthermore, the method for performing the screening analysis may be dependent upon the risk-informed application and the type of event being screened. Thus, it is appropriate to identify general criteria for the acceptability of a conservative or bounding analysis. Each conservative screening analysis must address the following to be acceptable:

- effects and completeness of potential impacts
- frequency of scenario
- appropriate to the risk-informed application

The spectrum of potential impacts of the missing scope or level-of-detail item and the effects on the evaluation of risk must be identified and addressed such that impacts or effects that could lead to a more severe credible outcome are not overlooked. That is, the different accident progressions have been identified and understood to the extent that a different and more severe credible outcome is unlikely.

For example, suppose that the PRA did not initially address LOCAs. If a conservative analysis were used to address LOCAs, the full spectrum of break sizes would need to be considered. If the spectrum only included break sizes from 8 inches to 24 inches, the analysis would not be bounding since the break sizes that were not considered (e.g., greater than 24 inches) could have a more severe outcome.

The frequency used in a conservative or bounding analysis should be greater than, or equal to, the maximum credible collective frequency of the spectrum of impacts analyzed for the missing item. Utilizing the LOCA example from above, if large LOCAs were missing from the PRA scope, a bounding assessment for this missing range of LOCAs would need to use the total frequency for the spectrum of LOCAs that are not in the PRA.

The screening process performed for a base PRA must also be reviewed for a risk-informed application to verify that the screening is still appropriate. This can be accomplished by confirming that the proposed plant modification or operational change does not affect the bases for screening performed in the base PRA. The cause-and-effect relationship used to establish the impact of the proposed change on SSCs and the required scope of the PRA discussed in Section 4.3 will provide the information necessary to accomplish this review. In this case, if there is no identified effect on specific scope items, then there will be no change in the risk metrics used to evaluate the plant change. However, if there is an identified effect, then either a screening or a detailed quantitative analysis (best-estimate or conservative) must be performed to determine the impact on the required risk metrics.

5.2.3 Substep C-1.3: Significance of Non-Modeled Scope Items

If a missing scope or level of detail PRA item is not screened using a bounding, conservative, or limited realistic analysis, the results of that analysis can be used in the application as a conservative risk estimate, if it is not significant to the risk-informed decision. A comparison of a conservative risk estimate against the appropriate risk acceptance guidelines for a risk-informed application will identify the significance of the non-modeled items. The degree to which the conservative risk estimate can be used to support the claim that the missing scope or level of detail in the PRA does not impact the decision depends on the proximity (quantitatively speaking) of the risk results to the guidelines. When the contributions from the modeled contributors result in a risk estimate that is close to the guideline boundaries, the argument that the contribution from the missing items is not significant must be more convincing than when the results are further away from the boundaries.

However, it is important to note that conservative risk estimates cannot be used in the same way a full PRA model is used to fully understand the contributions to risk thereby allowing the analyst to gain robust risk insights. Conservative risk assessments exaggerate the importance of initiating events, component failure modes, and accident sequences. As a result, the dominant contributors to risk may be masked, even though their contributions may be small. Thus, the usefulness of a conservative analysis is somewhat limited, particularly in applications relying on relative importance measures. The principal utility of a conservative analysis is to demonstrate that the risk contributions from non-modeled scope items as well as any change to the risk contribution that results from a change in the plant are small and, thus, not significant to the decision.

Care should be taken to make sure that any assumptions that are meant to screen out or bound a hazard group are not invalid for a specific application. For example, the assumption that tornados can be screened based on the assumption of the existence of tornado missile barriers, may be invalid for situations where a barrier is temporarily moved for a particular plant evolution.

5.3   Step C-2: Treatment of Non-Modeled Scope Items

The purpose of this step is to provide guidance for determining the possible ways to treat non-modeled scope or level-of-detail items that are significant to the decision. How a missing scope or level-of-detail item is treated is dependent on whether an NRC-endorsed standard exists that addresses the item.

The risk from each significant non-modeled scope or level-of-detail item should be addressed using a PRA model that is developed in accordance with a consensus standard for that item that has been endorsed by the NRC staff. As indicated in Section 4.3, a significant risk contributor is one whose inclusion in the application PRA model can impact the decision. PRA standards are developed to address specific hazards (e.g., internal, seismic, or high wind events) while a plant is in different plant operating states (POSs) (e.g., at-power, low-power, or shutdown). However, if there is no PRA standard that addresses the missing scope or level-of-detail item in question, the licensee can submit the results of the quantitative screening analysis as part of the input into the decisionmaking process. When this situation occurs, the analyst can then proceed to evaluate the parameter and model uncertainties in the context of the application.

If there is a consensus standard endorsed by the staff for a non-modeled scope or level-of-detail item that has been determined to be significant to the decision based on a screening evaluation, the licensee has several options:

- Upgrade the PRA model to include the missing scope or level-of-detail item. A detailed PRA model must be developed according to the endorsed standard.

- Redefine the application so that it does not affect the missing scope item (e.g., hazard or POS). When a PRA model does not completely cover all significant risk contributors (i.e., limited scope), the scope of implementation of a risk-informed change can be restricted to fall within the scope of the risk assessment. For example, if the PRA model does not address fires, the change to the plant could be limited such that it does not affect any SSCs that are used to mitigate the impact of fires. In this way, the contribution to risk from internal fires would be unchanged. This is the strategy adopted in Nuclear Energy Institute (NEI) report 00-04 [NEI, 2005b] for categorizing SSCs according to their risk significance when the PRA being used does not address certain hazard groups.

5.4   Summary of Stage C

This section provides guidance to the licensee for addressing the completeness in the PRA results that are used in support of risk-informed applications. Stage C is entered when it has been determined in Stage B that the existing PRA does not have the needed scope or contain the level of detail necessary for a specific risk-informed decision. Further the licensee has decided not to redefine the application or upgrade the PRA. The goal of this stage is to describe how to address the missing scope and level-of-detail items and, ultimately, determine whether those missing scope and level-of-detail items are significant to the decision under consideration.

Stage C contains guidance that is used to determine whether the missing scope or level-of-detail items can be screened using either qualitative or quantitative approaches. If the missing items can be screened, the licensee can proceed to Stage D. However, if a missing scope or level-of-detail item cannot be screened, the treatment of that missing item depends on whether it is addressed in an NRC-endorsed standard. If there is no PRA standard for the missing scope or level-of-detail item, the licensee can submit the results of a conservative analysis, bounding analysis, or both as part of the PRA evaluated in Stage D. If an endorsed PRA standard does exist, the risk from each significant scope item important to the decision should be addressed using a PRA model that is constructed and used in accordance with a consensus standard for the associated hazard or POS that has been endorsed by the staff. The resulting PRA is evaluated in Stage D. Alternatively, the licensee can redefine the application (i.e., returns to Stage B) such that it does not have any effect on the missing scope or level-of-detail item.

6.   STAGE D — ASSESSING PARAMETER UNCERTAINTY

This section provides guidance to the licensee on how to address the parameter uncertainties associated with the probabilistic risk assessment (PRA) that is used in support of risk-informed applications. This includes guidance on how to calculate the PRA results and the associated uncertainties that arise from the propagation of the underlying uncertainty in the input parameter values used to quantify the probabilities or frequencies of the events in the PRA logic model. In this step the determination is made whether the risk results challenge the quantitative acceptance guidelines, and whether the uncertainty in the results arising from the propagation of the underlying parameter uncertainty may be important for the comparison to the acceptance guidelines. The process of addressing parameter uncertainty corresponds to Stage D of the overall process for the treatment of uncertainties.

6.1   Overview of Stage D

In Stage B, the scope and level of detail of the PRA model needed to support the application has been defined and in Stage C any missing scope and level of detail has been addressed. At this stage, Stage D, the goal is to calculate the PRA risk metrics and determine how they compare to the quantitative acceptance guidelines, and if the uncertainty of the PRA risk results, due to the underlying parameter uncertainties, impacts this comparison. In order to properly assess the influence of the parameter uncertainty of the PRA inputs on the PRA results it is important to: (1) properly characterize the uncertainty in the parameters used in the various PRA inputs, (2) properly propagate that uncertainty through the analysis that calculates the risk metrics, while also properly accounting for the state-of-knowledge-correlation (SOKC) and its potential effect on the results, and (3) compare the results with the acceptance guidelines. Guidance for all three issues is provided in this section. This process involves the three steps illustrated in Figure 6-1. The steps are summarized here and discussed in detail in the following sections:

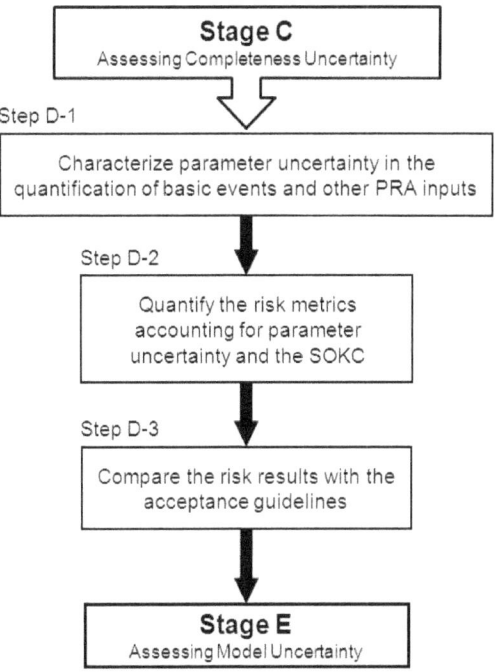

Figure 6-1 Overview of Stage D

- Step D-1: Characterize the uncertainty in the quantification of the parameters in the basic events[12] and other inputs of the PRA model. This characterization could, in the simplest approach, take the form of an interval (i.e., a range of probability values within which the actual input value lies). However, such a characterization will not lend itself to propagation of the parameter uncertainty through the PRA Therefore, it is more typical (and necessary if uncertainty propagation is to be achieved) to characterize the uncertainty in terms of a probability distribution on the value of the quantity of concern.

- Step D-2: Quantify the risk metrics, accounting for parameter uncertainty and the SOKC. This step involves quantifying the risk metrics, i.e., the output of the PRA, as well as estimating the uncertainty associated with the quantification. The uncertainty of the risk metric values is a consequence of the parameter uncertainty in the PRA input values. For the simplest approach, the risk metric can be quantified as a point estimate, in which case the uncertainty associated with the risk metric can only be expressed as an estimated uncertainty interval. For most applications ultimately requiring an NRC decision, the more desirable approach is to propagate the parameter uncertainties and calculate the risk metrics in such a way that the influence of the input parameter uncertainties on the mean value of the metrics, as well as a probability distribution reflecting the uncertainty about their mean value, is obtained. The effect of the SOKC on the results needs to be included in the calculation or, if neglected, justified as being negligible.

Step D-3: Compare the application risk results with the application acceptance guidelines. This step involves comparing the estimate of the relevant risk metric(s) with the acceptance guidelines to be used for the particular application being considered. This comparison reveals (1) if and how the acceptance guidelines are satisfied and, if needed for the decision, (2) how the uncertainty of the risk metric estimates, arising from the propagation of the uncertainty in the parameter values of the PRA inputs, impacts the comparison.

In providing guidance to the licensee for addressing parameter uncertainty, this section also provides guidance for meeting the NRC position[13] on the technical requirements on parameter uncertainty in the ASME and American Nuclear Society (ANS) PRA standard [ASME/ANS 2009]. The relevant requirements in the ASME/ANS PRA standard are related to characterizing the parameter uncertainty and to calculating the risk metrics, i.e., to Steps D-1 and D-2 in Figure 6-1. Steps D-1 and D-2 are addressed below in Sections 6.2 and 6.3, respectively. Therefore, the guidance of Steps D-1 and D-2 is provided in a context that also provides direction for meeting the NRC position on those requirements in the standard that relate to parameter uncertainty. The standard does not provide requirements regarding the comparison of the PRA results with acceptance guidelines. Consequently, the guidance provided for Step D-3 in Section 6.4 is not described in the context of the standard.

---

[12] The ASME/ANS PRA standard defines a basic event as an event in a fault tree model that requires no further development, because the appropriate limit of resolution has been reached.

[13] The NRC position is contained in an Interim Staff Guidance (ISG) document (to be published). All or some of the NRC position may be reflected in a future Addendum to the ASME/ANS PRA standard.

6.2 Step D-1 – Characterizing Parameter Uncertainty in the Quantification of Basic Events and Other PRA Inputs

The purpose of this step is to provide guidance on quantifying the probabilities (or frequencies) of the basic events and other PRA inputs that are formulated in terms of parameters with underlying uncertainty, and characterizing that uncertainty. The ultimate goal is to be able to calculate the mean and uncertainty of the risk metrics properly. In order to do that, one has to first correctly characterize the parameter uncertainty associated with the inputs to the PRA. It is especially important to capture the parameter uncertainty of those inputs that constitute significant contributors, i.e., those inputs that contribute significantly to the computed risk for a specific hazard group.

The ASME/ANS PRA standard defines a significant contributor in the following way, which is context dependent: (a) in the context of an internal events accident sequence/cutset, a significant contributor is a significant basic event or an initiating event that contributes to a significant sequence, (b) in the context of accident sequences/cutsets for hazard groups other than internal events, significant contributors also include the following: the hazard source, hazard intensity, and hazard damage scenario (for example, for Fire PRA, fire ignition source, physical analysis unit, or fire scenario that contributes to a significant accident sequence would also be included), (c) in the context of an accident progression sequence a significant contributor is a contributor that is an essential characteristic (e.g., containment failure mode, physical phenomena) of a significant accident progression sequence, and if not modeled would lead to the omission of the sequence.

As can be inferred from the above, significant basic events constitute a large subgroup among significant contributors. The ASME/ANS PRA standard defines a significant basic event as "a basic event that contributes significantly to the computed risks for a specific hazard group. For internal events, this includes any basic event that has a Fussell-Vesely (FV) importance greater than 0.005 or a Risk Achievement Worth (RAW) importance greater than 2. For hazard groups that are analyzed using methods and assumptions that can be demonstrated to be conservative or bounding, alternative numerical criteria may be more appropriate, and, if used, should be justified."

The definition of significant contributor also involves understanding what a significant accident sequence/cutset and significant accident progression sequence is. The ASME/ANS PRA standard defines a significant accident sequence as the following:

> one of the set of accident sequences resulting from the analysis of a specific hazard group, defined at the functional or systematic level, that, when rank-ordered by decreasing frequency, sum to a specified percentage of the core damage frequency for that hazard group, or that individually contribute more than a specified percentage of core damage frequency…(for the referenced version of the standard) the summed percentage is 95% and the individual percentage is 1% of the applicable hazard group…For hazard groups that are analyzed using methods and assumptions that can be demonstrated to be conservative or bounding, alternative numerical criteria may be more appropriate, and, if used, should be justified.

Similarly a significant cutset is defined as the following:

one of the set of cutsets resulting from the analysis of a specific hazard group that, when rank ordered by decreasing frequency, sum to a specified percentage of the core damage frequency (CDF) (or large early release frequency (LERF)) for that hazard group, or that individually contribute more than a specified percentage of CDF (or LERF)…(for the referenced version of the standard) the summed percentage is 95% and the individual percentage is 1% of the applicable hazard group. Cutset significance may be measured relative to overall CDF (or LERF) or relative to an individual accident sequence CDF (or LERF) of the applicable hazard group…For hazard groups that are analyzed using methods and assumptions that can be demonstrated to be conservative or bounding, alternative numerical criteria may be more appropriate, and, if used, should be justified."

Finally, a significant accident progression sequence is defined as the following:

one of the set of accident sequences contributing to large early release frequency resulting from the analysis of a specific hazard group that, when rank-ordered by decreasing frequency, sum to a specified percentage of the large early release frequency, or that individually contribute more than a specified percentage of large early release frequency for that hazard group…(for the referenced version of the standard) the summed percentage is 95% and the individual percentage is 1% of the applicable hazard group…For hazard groups that are analyzed using methods and assumptions that can be demonstrated to be conservative or bounding, alternative numerical criteria may be more appropriate, and, if used, should be justified. (Alternative criteria may be appropriate for specific applications. In particular, an alternative definition of "significant" may be appropriate for a given application where the results from PRA models for different hazard groups need to be combined.)"

In providing the guidance for parameter uncertainty, it is important to understand what is meant by the parameter uncertainty of a PRA input. It is defined as the uncertainty in the estimate of the probability or frequency of an input to the PRA due to the uncertainty in the parameter(s) associated with the corresponding mathematical model used to represent the input in the PRA. The choice of which model is used for a particular input can be a source of model uncertainty, as discussed in Section 7. The uncertainty in the parameters of the model is the subject of this section.

Characterization of the parameter uncertainty could, in the simplest approach, take the form of an interval (i.e., a range of probability values within which the actual input value lies). However, it is more typical to characterize the uncertainty in terms of a probability distribution on the value of the quantity of concern, for example a basic event parameter. In the case of basic events their probability (or frequency of an initiating event) is calculated using a probability model, which can be a function of one or more parameters, depending on the mathematical expression for the model. A simple example of a basic event model is the exponential distribution for the failure times of a component, which involves a single parameter, λ, the failure rate.

Since the guidance here is provided in the context of the NRC position on the pertinent ASME/ANS PRA standard requirements, it is also important to understand the relevant technical requirements (or supporting requirements, (SRs)) in the ASME/ANS PRA standard. The relevant SRs that define the types of PRA inputs[14] for which probabilities (or frequencies) need to be calculated, and parameter uncertainties characterized, are found throughout the parts of the standard that deal with the various internal and external hazards. For example, Part 2 of the standard, Requirements for Internal Events At-Power PRA, contains SRs regarding acceptable ways of dealing with parameter uncertainty when characterizing basic events used in the calculation of CDF. These include the following SRs for the PRA areas indicated:

- Initiating events for internal events: IE-C1, IE-C15
- Human failure events associated with internal events: HR-D6, HR-G8
- Component failures for internal events: DA-D1, DA-D3

In the LERF analysis (Section 2-2.8) of Part 2, SR LE-E1 refers back to the "CDF language" SRs for selecting parameter values and operator response for the accident progression, including the SRs above for parameter uncertainty.

Other Parts of the standard, which address other hazards, refer back to the SRs in Part 2 for carrying out analyses in initiating events, and component and human failure events, as appropriate. These other parts of the standard include requirements for analyzing the following hazards: internal flood (Part 3), fires (Part 4), seismic events (Part 5), high wind events (Part 7), external flood events (Part 8) and other hazard events (Part 9). (The standard currently lists requirements for conducting a PRA for at-power operation, i.e., requirements for a PRA for low power and shutdown operating states are not included.)

Furthermore, in the SRs for most of the hazards dealt with in Parts 3 through 9 of the standard, there are also additional mathematical models introduced (i.e., models other than the basic event models), that include parameters whose uncertainty must be considered. These models are used to characterize the hazard source, the hazard intensity, and the hazard damage scenario for the particular hazard being analyzed. For example, in carrying out a PRA for a fire hazard, parameter uncertainty must be considered in models used for the fire ignition source, fire propagation, and physical analysis units. For a seismic PRA, examples of additional inputs with parameter uncertainty are inputs used to characterize seismic sources, ground motion models, and fragility analysis. Therefore, there are unique additional SRs on characterizing parameter uncertainty found in these additional parts of the ASME/ANS PRA standard.

In developing the guidance for characterizing the uncertainty and quantifying the probabilities (or frequencies) of the PRA inputs, it should be recognized that the approach could vary, depending on the application. This variance is addressed in the standard by recognition that the level of detail, the level of plant specificity, and the level of realism needed in the PRA are commensurate with the intended application. Consequently, the standard defines three PRA Capability Categories (CCs) that are meant to support the range of applications.

The three CCs are distinguished by the extent to which:

- the plant design, maintenance, and operation are modeled
- plant-specific information with regard to SSC performance/history is incorporated
- realism with regard to plant response is addressed

---

[14] Some PRA inputs may not be applicable for a specific application. The applicable inputs are determined in Stage B.

Generally, from CC I to CC III, the extent to which the level of detail, plant specificity, and realism are modeled in the PRA increases. However, not every SR has distinct requirements defined for each CC. What is needed for a specific SR may not vary across the CCs; however, there may be variance in the implementation of the requirement. For example, the parameter uncertainties need to be characterized; however, the approach used may vary.

Since the characterization of the uncertainty for basic events is carried out similarly in all parts of the standard, more detailed guidance on the NRC position on the treatment in the standard of basic event parameter uncertainty is presented here. The treatment of other PRA inputs with parameters whose uncertainty must be characterized is carried out in an analogous manner. As noted above, the requirements for the calculation of the probabilities (and frequencies) of the basic events, and the characterization of the associated parameter uncertainty, is dependent on the capability category (CC) in the standard. This is true also of the NRC position on these requirements, which is the following:

- Capability Category I: Point estimates of the basic event parameters are calculated and their uncertainty is characterized qualitatively. This qualitative characterization involves specifying a range of values (an uncertainty interval) or by giving a "qualitative discussion" of the range of uncertainty.

- Capability Category II: Mean values are calculated for the parameters of the **significant** basic events, and the uncertainty is characterized by providing a probabilistic representation of the uncertainty of the parameter values for the **significant** basic events (see the definition of a significant basic event provided above in this section). This characterization will enable the evaluation of a mean value and will provide a statistical representation of the uncertainty in the value of the parameter. Moreover, this characterization will allow the propagation of this uncertainty characterization through the PRA model so that the mean value of the PRA results (e.g., CDF, LERF, accident sequence frequency) and a quantitative characterization of their uncertainty can be generated.

- Capability Category III: Same as CC II but mean values are calculated for the parameters of **all** the basic events, and the uncertainty is characterized by providing a probabilistic representation of the uncertainty of the parameter values for **all** the basic events.

Acceptable approaches for meeting the above capability categories include the following:

Capability Category I

In this CC, an acceptable method for calculating the point estimates of the basic events includes the use of applicable generic data, since this category requires less than the other categories for level of detail, level of plant specificity, and level of realism. However, plant-specific information is still necessary for unique basic events, i.e., basic events involving unique initiating events, unique SSCs, or unique human actions. Generic data is used when the quality or quantity of plant-specific data is insufficient. For example, when a PRA is being performed on a new plant that has no operating history or when no plant-specific information exists for a specific component, an existing generic data base may have to be used to obtain information about the uncertainty of a parameter. In NUREG/CR-6823, "*Handbook of Parameter Estimation for Probabilistic Risk Assessment*" [SNL, 2003], Section 4.2 lists several applicable generic data sources that are currently available and used throughout the nuclear power PRA industry. This section of the handbook includes generic data bases sponsored by the Department of Energy (DOE) for use in PRAs, generic data bases developed by organizations related to the nuclear-power industry, a summary of two foreign data bases (from Nordic nuclear power plants (NPPs)), and a discussion of several non-nuclear data bases, which could be useful for some data issues in NPP PRAs. Additionally, Section 4.2.6 of the Handbook gives several cautions when using this type of data base.

For CC I, an acceptable method for characterizing the uncertainty of the basic event parameters can be a qualitative characterization. This qualitative characterization specifies an uncertainty interval, i.e., an upper and lower bound of probability values within which the actual basic event probability value lies; or can involve a qualitative discussion of the range of uncertainty. For example, the uncertainty range can be expressed in terms of factors on the point estimate.

Capability Category II

An acceptable method for this CC requires that, for the **significant** basic events (see the definition of a significant basic event provided above in this section), mean values of the basic event parameters are calculated and the uncertainty of the parameters is characterized by providing a probabilistic representation of the uncertainty of the parameter values. For non-significant basic events, the parameters could be point estimates, for example screening values with no distributions, such as screening values for human error probabilities (HEPs). In this CC, for the significant basic events, acceptable methods for calculating the mean values of the basic event parameters, as described below, are (1) Bayesian updating and (2) expert judgment. The uncertainty of the parameters is characterized by providing a probabilistic representation of the uncertainty of the parameter values. (Frequentist methods are not used because they do not produce probabilistic representation of the uncertainty.)

- Bayesian Approach. The Bayesian approach characterizes what is known about the parameter in terms of a probability distribution that measures the current state of belief in the possible values of the parameter. The mean value of this distribution is typically used as the point estimate for the parameter. The Bayesian approach provides a formal approach for combining different data sources. Details on the Bayesian Approach can be found in a number of statistical texts and in NUREG/CR-6823 [SNL, 2003].

- Expert Judgment. The expert judgment approach relies on the knowledge of experts in the specific technical field who arrive at best estimates of the distribution of the probability of a parameter or basic event. This approach is typically used when the needed information is very limited or unavailable. Such a situation is usual in studying rare events. Ideally, this approach provides a mathematical probability distribution with values of a central tendency of the distribution (viz., the mean) and of the dispersion of the distribution, such as the 5th and 95th percentiles. The distribution represents the expert or "best available" knowledge about the probability of the parameter or basic event. The process of obtaining these estimates is typically called "expert judgment elicitation," or simply "expert judgment" or "expert elicitation." For expert judgment, details can be found in NUREG/CR-6372 [LLNL, 1997] or NUREG-1563 [NRC, 1996].[15]

Capability Category III

---

[15] In addition, NUREG/CR-4550, Vol. 2 provides the approach used in NUREG-1150 on expert elicitation.

An acceptable method for this CC has the same requirements as that for CC II, except that for CC III parameters for all the basic events have their mean values calculated and the uncertainty of the parameters characterized by a probabilistic representation of the uncertainty of the parameter values. As for CC II, for CC IIII acceptable methods for calculating the mean values of the basic event parameters are (1) Bayesian updating and (2) expert judgment.

Conceptually, the guidance provided in this section applies to all the types of basic events (i.e., initiating event frequencies, component failure probabilities and HEPs). The following references provide more specific guidance for CCFs and human failure events:

CCFs of Components[16]

- NUREG/CR-6268 [INL, 1998]
- NUREG/CR-4780 [EPRI, 1988]
- EPRI NP-3967 [EPRI, 1985]

Human Failure Events

- NUREG-1792 [NRC, 2005]

For convenience the above discussion on what is acceptable for the three CCs in the ASME/ANS PRA standard focused on basic events. However, it is also applicable to other PRA inputs, needed in addition to basic events, when other hazards besides the internal event hazard is considered (for example seismic hazard intensity), or accident progression beyond core damage is considered (for example probability of a containment failure mode). The differentiation between CCs II and III is then in terms of all significant contributors, not just significant basic events (see the definition of a significant contributor provided above in this section). Additional guidance on the characterization of hazard specific PRA contributors can be found in the reference sections of the ASME/ANS PRA standard part applicable to the specific hazard.

In characterizing the uncertainty and quantifying the probabilities (and frequencies) of the PRA inputs, there are a few issues of concern for some special basic events and other contributors that need to be addressed.

- There are reference documents that explicitly address the estimates of the probabilities of failures to recover SSCs and the probabilities of special events (e.g., the probability of reactor coolant pump seal failure). These documents should also explicitly address the parameter uncertainty associated with the particular event. In some cases, however, the analysis provides an estimate of the probability of an event without giving the associated probability distribution. For example, the uncertainties in the probabilities of recovering alternating current (ac) power after a loss of offsite power are sometimes not evaluated. In these cases, an estimate of the probability distributions of recovery should be included in the basic event characterization so the parameter uncertainty associated with a particular event is included in the PRA model. There should be some justification for the estimate of the probability distribution provided. The type of justification will depend on the basic event modeled and may need to reference data from a similar plant or location, or generic or plant-specific calculations.

---

[16] NUREG/CR-5497 and NUREG/CR-5485 contain additional guidance on the treatment of CCF for components.

- The probabilities of failing to recover SSCs and the probabilities of special events can be derived from models, which are quite complex. For instance, in the fire PRA, the probability of a fire scenario is obtained from the combination of models of fire growth and suppression phenomena. As noted above, this fire scenario probability should have an associated parameter uncertainty. (This parameter uncertainty should not be confused with the uncertainty related to the choice of model, which is discussed in Section 7 of this document.)

It should also be noted that the parameter uncertainty in PRA models of new reactors (new light water reactors (LWRs), advanced light-water reactors (ALWRs) and especially non-LWR advanced reactors) is expected to be greater than for currently operating LWRs. Greater parameter uncertainty is likely because of the lack of operational history, and, therefore, lack of data for some components of these new reactors. This lack of data will influence the approaches that are discussed here. For example:

- Lack of plant specific operating experience and data - Plant-specific operating experience will not be available. An alternative is to use generic data, and plant specific operating experience from existing plants, as applicable.

- Lack of applicability of generic data - Generic data may not be applicable for some of the ALWR and non-LWR advanced reactor design features. Generic nuclear data may not be available if equipment has not previously been used in nuclear plant applications. An alternative action is to evaluate the applicability of other generic data to the new plants under consideration. Expert judgment can be used to extrapolate from both nuclear and non-nuclear data when the applicability and/or availability of the nuclear data are limited.

- For both these issues the related assumption is that non-nuclear data can be used to help establish a basis for the performance of nuclear SSCs that share a similar technical pedigree to non-nuclear SSCs with existing data. Furthermore, expert judgment can be used to carefully assess the impact of extenuating circumstances (e.g., high temperatures, new materials).

At this point in Stage D, the parameters of the essential PRA inputs have been quantified and their parameter uncertainty characterized. Assuming the appropriate requirements of the ASME/ANS PRA standard as endorsed by NRC have been met, it is now possible to calculate the risk metrics, and characterize their uncertainty due to the parameter uncertainty of the PRA inputs. Once the risk metrics are calculated it will be possible to determine if, and to what degree, the acceptance guidelines are met, challenged, or exceeded.

6.3   Step D-2 – Quantifying the Risk Metrics while Accounting for Parameter Uncertainty and the SOKC

The purpose of this step is to provide guidance for quantifying the frequencies and/or probabilities of the risk metrics and estimating their uncertainty, i.e., the uncertainty of the risk metric due to propagation of parameter uncertainties through the PRA. Estimating the uncertainty of the risk metric resulting from propagating the parameter uncertainty could, in the simplest approach, i.e., used in CC I, take the form of an interval (e.g., a range of estimates within which the actual risk metric value lies). However, it is more typical to characterize the uncertainty in terms of a probability distribution on the value of the quantity of concern. For CCs

II and III, the mean and the distribution for the risk metric estimates are usually obtained by propagating the parameter uncertainties of the PRA inputs through the analysis using the Monte Carlo or similar sampling method. The difference between CC II and CC III is that in CC II the propagation of the uncertainty is only carried out for significant contributors (see the definition of a significant contributor provided above in this section) in the significant accident sequences and cutsets (see the definition of a significant accident sequence/cutset and significant accident progression sequence provided above in this section), while for CC III the uncertainty distribution for all the input parameters is propagated to obtain the mean of the risk metrics as well as their uncertainty distributions.

In carrying out the propagation, it is important to consider the state of knowledge correlation (SOKC) between events. The SOKC arises because, for identical or similar components, the state-of-knowledge about their failure parameters is the same. In other words, the data used to obtain mean values and uncertainties of the parameters in the basic event models of these components may come from a common source and, therefore, are not independent, but are correlated. When the basic event mean values and uncertainty distributions are propagated in the PRA model without accounting for the SOKC, the calculated mean value of the relevant risk metric and the uncertainty about this mean value will be underestimated. The values can be underestimated due to the effect of the SOKC directly, as well as due to incorrect screening out of cutsets in truncation due to neglect of the SOKC in calculating cutset frequencies. Appendix 6-A of this section discusses both these potential effects of the SOKC in more detail. The influence or importance of the SOKC on the value of the risk metrics will vary from case to case.

In most risk-informed applications the risk metrics of concern are CDF and LERF. Since the guidance in this step is also provided in the context of the ASME/ANS PRA standard, as modified by the NRC position, it is important here also to understand the relevant SRs in the ASME/ANS PRA standard. Part 2 of the standard, contains the SRs for internal events at-power, and includes the SRs delineating acceptable ways of dealing with the propagation of parameter uncertainty from basic events when calculating CDF. These include QU-A3 for calculating CDF and QU-E3 for estimating the uncertainty in CDF. The LERF calculation of Part 2 refers back to the applicable SRs in the CDF calculation for calculating LERF and estimating its uncertainty.

SRs in the other parts of the standard also refer back to the Part 2 SRs, but also supplement these with additional SRs, where needed, for propagation of all the parameter uncertainty (not just the parameter uncertainty in the basic event models) in the calculation of the risk metrics and their uncertainty.

Similar to Step D-1, where the CC played a role in the characterization of the parameter uncertainty of the basic events, the calculation of the risk metrics and characterization of their associated parameter uncertainty is also dependent on the CC, as described below. However, regardless of the CC, it is necessary to determine if the SOKC is important in significant sequences and/or cutsets.

- Capability Category I: When the SOKC is unimportant in significant sequences/cutsets a point estimate is calculated for the risk metric. When addressing the uncertainty interval or probability distribution, an estimate of the uncertainty interval and its basis is sufficient. If the SOKC is important in significant sequence/cutsets, the calculation of the risk metric and the characterization of its associated parameter uncertainty is carried out to meet CC II requirements.

- Capability Category II: If the SOKC is important in significant sequence/cutsets, a mean[17] value is calculated for the risk metric by propagating the uncertainty distributions for the significant contributors through all significant accident sequences/cutsets using the Monte Carlo approach (or other comparable means) through the PRA model, ensuring that the SOKC between event frequencies or probabilities is taken into account. The uncertainty distribution of the risk metric is calculated by propagating the uncertainty distributions of the significant contributors through all significant sequences/cutsets using the Monte Carlo or similar approach and taking the SOKC into account. If the SOKC is not important in significant sequence/cutsets, a mean value is calculated for the risk metric using the mean values of significant contributors, and the uncertainty interval of the risk metric can be estimated taking into account the uncertainty distributions of the significant contributors to the risk metric.[18]

- Capability Category III: A mean value is calculated for the risk metric by propagating the uncertainty distributions of all the input parameters (both significant and non-significant contributors) using the Mont Carlo approach (or other comparable means) through the PRA model, ensuring that the SOKC between event frequencies or probabilities is taken into account. The uncertainty distribution of the risk metrics is calculated by propagating the uncertainty distributions of all the contributors through all retained sequences/cutsets using the Monte Carlo or similar approach and taking the SOKC into account.

Before providing more detailed guidance on how to meet the individual CCs, guidance on determining the importance of the SOKC is also needed. Depending on the methods and software used in carrying out the PRA calculations, it is often simpler to just account for the SOKC in the PRA model rather than to first try to establish its importance beforehand. In other words, establishing the SOKC's importance may be just as, or more, cumbersome than performing the PRA calculation including the SOKC. However, although accounting for the SOKC is the preferred approach, there may be situations where this is difficult for the licensee, for example if there is an existing PRA that was quantified without accounting for the SOKC.

> For such conditions the Electric Power Research Institute (EPRI) has formulated some guidelines for ascertaining the importance of the SOKC.
>
> Section 2.4, "Guidelines for Addressing Parametric Uncertainty," of EPRI report 1016737 [EPRI, 2008] presents some guidelines for addressing the SOKC in evaluating the value of a risk metric and the uncertainty of the risk metric that results from the parameter uncertainty of the PRA inputs, i.e., for meeting the QU-A3 and QU-E3 SRs of the ASME/ANS PRA standard.

Separate guidelines are presented for meeting the SRs when dealing with (1) the base PRA model or (2) a PRA application:

- Two guidelines, each split into a preferred and alternate approach, are presented for calculating mean values for CDF and LERF, the first guideline is for the base model and

---

[17] This is actually an approximation of the true mean since only significant contributors in significant sequences/cutsets are included.

[18] This is actually an approximation of the true mean since only significant contributors in significant sequences/cutsets are included and the SOKC is not taken into account.

the second for applications. For both the base model case and the application case the EPRI guidelines affirm that the preferred approach for obtaining the mean value of the risk metric is to ensure that the SOKC is appropriately represented for all relevant events and to perform a detailed Monte Carlo (or similar) calculation with enough samples to demonstrate convergence. For the base model, if the preferred approach of appropriately accounting for the SOKC cannot be completed, an alternate approach involving comparison with a PRA, modeling similar features that has evaluated the effect of the SOKC is suggested. The difficulties with this approach are acknowledged. For an application an alternative is suggested where the cutsets of the PRA model of the application are reviewed to establish whether the risk metric used for the application is determined by cutsets that involve basic events with SOKC correlations. If they do not, then the point estimate of the risk metric can be used instead of the mean value. However, the guidelines acknowledge that this may not be practical to implement because it would require a detailed review of cutsets that have an impact on the risk metric to determine if basic events that are correlated are present.

- Two additional guidelines, one for the base model and one for the application, each again split into two alternatives, are offered if the risk metric uncertainty arising from the uncertainty of the PRA parameter inputs has to be provided for decisionmaking. For the base model the preferred approach to obtain the risk metric uncertainty is again to perform parametric uncertainty propagation on the PRA model using a Monte Carlo or similar process through the cutsets, accounting for the SOKC. The alternative is again a comparison with an existing PRA that has evaluated the parametric uncertainty taking into account the SOKC. The difficulties with this approach are again acknowledged. For an application the first alternative calls for demonstrating that the probability distribution is not expected to significantly change (e.g., because the significant contributors for the application do not involve correlated basic events) from the base-model probability distribution. If this condition is satisfied, the base-model probability distribution is used for the application. If this cannot be demonstrated, the alternative is to appropriately account for the SOKC in evaluating the parameter uncertainty of the risk metric of the PRA application by setting up the groups of basic events that are correlated in the model and propagating the parameter uncertainty in the model.

Acceptable approaches for meeting the SRs in the ASME/ANS PRA standard related to calculating the risk metrics while accounting for parameter uncertainty can now be discussed for each of the three capability categories of the standard. However, regardless of the CC, it is necessary to determine if the SOKC is important in significant sequences and/or cutsets. In describing these approaches, the order in which the analysis is performed does not necessarily follow the order presented here. Moreover, in the subsequent discussion it is assumed that a PRA model was implemented using a PRA computer code.

Capability Category I

For CC I acceptable methods for characterizing uncertainty in the quantification of the risk metrics, due to uncertainty in the input parameters, contain the following components when the SOKC is not important:

Evaluation of the PRA model to generate the point estimate of a risk metric. For CC I SR QU-A3 can be satisfied with a point estimate of CDF or LERF. The solution of the PRA model yields the cutsets of the PRA logic model. The PRA computer code then generates a point estimate of the CDF or LERF by quantifying these cutsets. The code uses the point estimates of the basic events to obtain the point estimate of the CDF or LERF. (For PRA methodologies that do not use a 'cutset' methodology point estimates are calculated as appropriate for the methodology.)

Estimation of the uncertainty interval of the risk metric. For CC I, SR QU-E3 can be satisfied by providing an estimate of the uncertainty interval of the risk metric and a basis for the estimate, consistent with the characterization of parameter uncertainties. The risk metric uncertainty estimation and its basis are case-specific and, for this reason, no general guidance is offered.

If the SOKC is important in significant sequence/ cutsets, the calculation of the risk metric and the characterization of its associated parameter uncertainty is carried out to meet CC II requirements.

Capability Category II when the SOKC is not important

As previously noted, establishing whether the SOKC is important may be as difficult, or more so, as simply appropriately including the SOKC when performing the PRA calculation. If it can be established that the SOKC is not important the following considerations apply.

> For some cases the guidelines of EPRI report 1016737 [EPRI, 2008] may be used for ascertaining the importance of the SOKC prior to completing the PRA quantification.

In this case SR QU-A3 for CC II can be satisfied by calculating the mean[19] of the risk metric using the mean values of the significant contributors (see the definition of a significant contributor provided above in this section) in the significant accident sequences/cutsets (see the definition of a significant accident sequence/cutset and significant accident progression sequence provided above in this section). In other words, once the cut sets of the logic model of the PRA have been established, the mean values of the significant contributors are used in the PRA computer code and propagated through the significant accident sequences/cutsets to obtain the mean values of the CDF or LERF. The mean value of each significant contributor means the mean value of the probability distribution of that contributor. Note that, in light of the definition of significant accident sequence/cutset, this approach may capture only 95% of the CDF or LERF. This is considered sufficient by NRC to satisfy this SR for CCII.

To satisfy SR QU-E3 for CC II when the SOKC is not important, the uncertainty interval of the CDF and/or LERF results can be estimated by taking into account the uncertainty distributions of significant contributors and those model uncertainties characterized by a probability distribution.

Capability Category II when the SOKC is important and Capability Category III

For meeting CC II when the SOKC is important and for meeting CC III regardless of the importance of the SOKC, acceptable methods for characterizing parameter uncertainty in the quantification of the risk metrics for CC II involve the following five substeps. The difference between CC II and CC III is that in CC II the propagation of the uncertainty is only done for significant contributors (see the definition of a significant contributor provided above in this section) in the significant accident sequences and cutsets (see the definition of a significant

---

[19] This is actually an approximation of the true mean, since only significant contributors in significant sequences/cutsets are included and the SOKC is not taken into account.

accident sequence/cutset and significant accident progression sequence provided above in this section), while for CC III the uncertainty distribution for all the input parameters is propagated to obtain the mean[20] of the risk metrics as well as their uncertainty distributions.

1. *The PRA model is evaluated to generate the point estimate of the risk metric.* The solution of the PRA model yields the cut sets of the logic model of the PRA. The PRA computer code generates a point estimate of the CDF or LERF by quantifying these cut sets. The code uses the point estimates of the inputs to obtain the point estimate of the CDF or LERF. The point estimate of each PRA input that contains parameter uncertainty must equal the mean value of the probability distribution of that basic event. The cut sets obtained from this evaluation procedure are subsequently used to propagate parameter uncertainties throughout the logic model of the PRA, as described in Substep 5. It should be noted here that, due to the large number of cutsets in a PRA model, it is common to screen out cutsets with frequencies below a certain truncation value at this point in the analysis. Caution needs to be exercised to avoid incorrect screening out of cutsets in truncation due to neglect of the SOKC in calculating their frequencies. Appendix 6-A of this section discusses this possible effect of the SOKC in more detail, along with other potential SOKC effects.

2. *The parameter uncertainty data for each basic event is entered into the PRA model.* For each input to the PRA that contains parameter uncertainty, the information about the probability distribution of each of its parameters must be entered into the PRA code. For example, if a basic event model is an exponential distribution with parameter λ, data about the distribution of this parameter are entered into the code. The distributions of the parameters of all these inputs are subsequently used to propagate parameter uncertainties through the PRA model.

3. *The groups of basic events that are correlated, due to the SOKC, are defined.* When evaluating the PRA model to calculate a risk metric or an intermediate value, such as the frequency of an accident sequence, the correlation between the estimates of the parameters of some basic events of the model, the SOKC, must be taken into account for all significant basic events for CC II and for all basic events for CC III. (Appendix 6-A to this section discusses the fundamental principles of the SOKC.).

    The first step in accounting for the SOKC between basic events is identifying those events that are correlated and grouping them. Each identified group contains basic events that are correlated with each other because the analysts' state-of-knowledge about the parameters for these events is the same.

    Correlated events are identified by determining which basic event models share the same parameters, i.e., are quantified from the same data. For example, when considering all components of a certain type in a NPP, if the failure rate for that component type's failure mode is evaluated from the same data set, the basic events for these components are correlated. However, all the components of a certain type in a NPP do not have to be correlated. For example, if the failure rates for subgroups are determined using different data sets, the basic events for these components are correlated within the subgroups, but not across the subgroups. Accordingly, for a particular PRA model, several different groups of correlated basic events can be defined.

---

[20] Therefore, for CC II this is actually an approximation of the true mean, since only significant contributors in significant sequences/cutsets are included.

The groups of basic events correlated via the SOKC should not be confused with groups of common cause failures (CCFs). Although both groups account for statistical correlations between the estimates for component failure of a NPP, they account for different correlations. For this reason, accounting for one type of correlation does not account for the other. A group of correlated basic events can contain several events, including those modeled within a CCF group. For instance, a CCF group may contain one failure mode of all the pumps of a particular system, while a group of correlated basic events may encompass the same failure mode for all the pumps of this type within the NPP. Hence, both types of correlations (i.e., CCF and SOKC) should be included in a PRA model.

4. *Each group of correlated basic events is appropriately entered into the evaluation code.* Each group of correlated basic events (defined in the third step) in the PRA model should be set up in a PRA computer code such that the particular code recognizes that the basic events are correlated, i.e., that their uncertainty is characterized by the same probability distribution. In this way, a single distribution is used to model the uncertainty of each basic event in a correlated group. Then, when the code propagates the uncertainty, for each sampling run through the PRA model, the same sample from the correlated group's distribution is used to quantify each basic event in the group. These values of the basic events are subsequently used in propagating parameter uncertainties through the PRA model to generate a value of the risk metric of interest, such as the CDF. This evaluation process is repeated for all the samples evaluated by the code.

5. *The calculation of the risk metrics, and their uncertainty evaluation, is carried out by propagating the parameter uncertainties of all applicable PRA inputs through the PRA model.* An uncertainty evaluation of the PRA model must be performed by executing an uncertainty calculation of the model in the PRA computer code to obtain a risk metric estimate, such as CDF. The PRA model was previously set up in the code according to Steps 1 through 4, as applicable. Provided that the PRA model was set up such that the code recognizes that some events are correlated (as described in Step 4), the code will automatically account for the SOKC when running the uncertainty evaluation unless a specific code requires additional steps.

The uncertainty can be evaluated using the Monte Carlo or Latin Hypercube Sampling (LHS) methods. The number of samples used should be large enough so that the sampling distribution obtained converges to the true distribution of the risk metric. The standard error of the mean (SEM) is a measure of this convergence and is shown below in equation 6-1.

$$SEM = \frac{\sigma}{\sqrt{n}}$$
Equation 6-1

In this equation, σ is the standard deviation of the sampling distribution (i.e., the square root of the variance of this distribution) and n is the number of trials. Evaluation of the above equation demonstrates that using a larger number of samples will produce more accurate estimates of the sampling mean.

An iterative process is required in which several consecutive uncertainty calculations are executed, each with an increasing number of samples. At the end of each run, the SEM is calculated. The process may be stopped when (1) increasing the number of samples does not significantly change the SEM or (2) the SEM reaches a predefined small error value.

It is advisable, though unnecessary to use the LHS method, rather than the Monte Carlo approach since the LHS requires a significantly smaller number of samples to ensure a robust sampling of distributions as compared to the Monte Carlo approach.

The result of the uncertainty evaluation of the PRA model is a mean value and the uncertainty distribution of the risk metric of interest, such as the CDF or LERF.

6.4 Step D-3 – Comparing the Risk Results with the Application Acceptance Guidelines

The purpose of this step is to provide guidance for comparing the PRA results with acceptance guidelines. In this step the determination is made whether the risk results challenge the quantitative acceptance guidelines, and whether the uncertainty in the results arising from the propagation of the underlying parameter uncertainty may be important for the comparison to the acceptance guidelines.

To make this determination (i.e., compare the results to the acceptance guidelines), the information needed consists of an estimate of the relevant risk metric(s), usually expressed as the mean value(s), and the acceptance guidelines to be used for the particular application. For some cases the uncertainty interval or distribution(s) of these risk metric(s) is also of interest.

Using this information, it is possible to understand if, and with what margin, the acceptance guidelines are satisfied by the risk results, given the parameter uncertainties.

Mean values for the risk metric estimates are usually the key results from the risk analysis because most acceptance guidelines are currently, either explicitly or implicitly, expressed in terms of the mean values of a risk metric. An approach using mean values is conceptually simple and consistent with classical decisionmaking. Moreover, an evaluation of the mean value incorporates a consideration of those uncertainties explicitly captured in the model. Since the mean is the average value obtained from a probability distribution, its value depends on that distribution, and, therefore, is representative of, in some measure, the uncertainty denoted by the distribution[21]. However, it should be kept in mind that the mean, as any other single estimate derived from a distribution, is a summary measure that does not fully account for the information content of the probability distribution.

These ideas are elaborated in a SECY paper issued by NRC in December of 1997 (SECY-97-287), whose subject was "Final Regulatory Guidance on Risk-Informed Regulation: Policy Issues." The SECY paper recommends that parametric uncertainty (and any explicit model uncertainties) in the assessment be addressed using mean values for comparison with acceptance guidelines. The mean value (or other appropriate point estimate if it is arguably close enough to the mean value) is appropriate for comparing with the acceptance guidelines. This approach has the major advantage of being consistent with the current usage of acceptance guidelines, (i.e., the Commission's Safety Goals and subsidiary objectives are meant to be compared with mean values). The SECY paper also points out that for the distributions generated in typical PRAs, the mean values typically corresponded to the region of the 70th to 80th percentiles. Coupled with a sensitivity analysis that is focused on the most

---

[21] The median, on the other hand, is simply the middle value in a probability distribution and, therefore, contains less information about the distribution than the mean.

important contributors to uncertainty, these mean values can be used for effective decisionmaking[22].

The specific acceptance guidelines and their form depend on the particular application being considered. The form of the acceptance guidelines also plays a role in determining the appropriate uncertainty comparison. For example, the acceptance guidelines in Regulatory Guide 1.174 [NRC, 2002], which provides an approach for using PRA in risk-informed decisions on plant-specific changes to the licensing basis, require the risk metric estimates of CDF and LERF and their incremental values, ΔCDF and ΔLERF, to be compared against a figure of acceptable values. In this example, the means of the risk metrics and the means of their incremental values need to be established for comparison with the figure of acceptable values.

The actual information presented in the application regarding the risk metric estimate and its uncertainty again depends on the ASME/ANS PRA standard CC that the pertinent SRs in the PRA model are intended to meet.

Capability Category I

For CC I, only the point estimate of the risk metric is required when comparing with the relevant acceptance guidelines. If the point estimate is not the mean this is problematic for many applications. At a minimum, the estimate of the uncertainty in the metric must be expressed in terms of an uncertainty interval, along with a basis for the provided interval.

Capability Category II or III

For CC II, or CC III, the standard requires providing the mean value for a risk metric. For CC II, if the SOKC is not important, a mean of the risk metric can be obtained by propagating the means of the significant contributors (not a true mean since non-significant contributors are neglected). For CC II, when the SOKC is important, and always for CC III, the mean of the risk metric is obtained by propagating parameter uncertainties through the PRA model using a Monte Carlo or equivalent sampling method. The difference between CC II and CC III is that in CC II the propagation of the uncertainty is only carried out for significant contributors (see the definition of a significant contributor provided above in this section) in the significant accident sequences and cutsets (see the definition of a significant accident sequence/cutset and significant accident progression sequence provided above in this section), while for CC III the uncertainty distribution for all the input parameters is propagated to obtain the mean[23] of the risk metrics and their uncertainty distributions.

Regardless of the CC, the comparison of the risk metric estimates to the acceptance guidelines should demonstrate to the decisionmaker whether the guidelines have been met or not, as well as the proximity of the risk results to the guidelines. The uncertainty of the risk metric estimate resulting from the propagation of the parameter uncertainty may also be of interest to the decisionmaker. If the risk metric mean value (or point estimate for CC I) meets the acceptance guidelines with plenty of margin, the uncertainty may not be important for the decision. On the

---

[22] The SECY paper recommends using sensitivity studies to evaluate the impact of using alternate models for the principal implicit model uncertainties. To address incompleteness, the SECY paper advocates employing quantitative or qualitative analyses as necessary and appropriate to the decision and to the acceptance guidelines.

[23] Therefore, for CC II this is actually an approximation of the true mean, since only significant contributors in significant sequences/cutsets are included.

other hand, if the risk metric estimate "challenges" the acceptance guideline, i.e., comes closer to the maximum allowable value, the uncertainty distribution (or range) is likely to play a more important role in the eventual decision, since it will provide some information on how likely it is that the acceptance values could be exceeded. Similarly, if the mean or point estimate value of the risk metric estimate exceeds the acceptable value by a small amount, the uncertainty information can provide insights on the importance of the excess.

6.5   Summary of Stage D

This section provides guidance to the licensee for addressing the quantification of the PRA results that are used in support of a risk-informed application, while accounting for parameter uncertainty. The ultimate purpose here is to determine whether (and the degree to which) the mean value PRA results and their uncertainty (from the propagation of the underlying parameter uncertainties) compare with the quantitative acceptance guidelines. To accomplish this purpose the guidance presented involves characterization of parameter uncertainty, propagation of parameter uncertainties (which includes an assessment of the significance of the SOKC), and comparison of results with acceptance guidelines.

However, parameter uncertainty is only one type of uncertainty in the PRA results that needs to be considered for risk-informed decisionmaking. Therefore, with the completion of Step D, the overall process now proceeds to Step E, the treatment of modeling uncertainties.

APPENDIX 6-A: THE STATE-OF-KNOWLEDGE CORRELATION

6-A.1 Definition of the State-of-Knowledge Correlation

Many of the acceptance guidelines used in risk-informed decisionmaking are defined such that the appropriate measure for comparison with the acceptance guidelines is the mean value of the uncertainty distribution of the relevant risk metric estimate. In calculating the mean value and uncertainty distribution of the risk metric estimate, it is important to understand the impact of a potential correlation among input parameters, referred to as the state-of-knowledge correlation (SOKC). The purpose of this appendix is to provide an explanation of the SOKC and its possible effect on the mean value and uncertainty distribution of the risk metric estimate. As explained below, the SOKC stems from the fact that, for identical or similar components in a given nuclear power plant, the state of knowledge about their failure parameters is the same. Apostolakis and Kaplan [Apostolakis 1981] described this correlation, and parts of this discussion are based on their paper, as well as discussion in EPRI Report 1016737 [EPRI, 2008].

In general, the input parameters used to quantify the basic events in the probabilistic risk assessment (PRA) model are represented by probability distributions representing the parametric uncertainty in those parameter values. In other words, the PRA basic event parameter mean values should be equal to the mean values of these distributions developed from the generic or plant specific component failure data base or operating experience.

In the ideal situation, each plant initiating event, structure, system, or component or operator action modeled as a basic event in the PRA would have its own database associated with it and thus would be statistically independent (i.e., their parameter values would be based on independent data that is not pooled or correlated in any way). If this were the case, the propagation of the basic event mean values in the analysis would lead to point estimates of the risk metrics that would themselves be true mean values. However, in general, this ideal situation is not realized in practice, and the data used for like components within a cut set of the analysis often has some common element, is pooled, or is correlated in some way. For example, the generic knowledge of the failure rate of one particular pump (such as a low-pressure coolant injection pump) for a given failure mode is typically based on experience with all "similar" pumps. Therefore, the various basic events that involve this failure mode of a pump are all in fact being estimated from a single state-of-knowledge distribution, and the data used for the pumps is not independent but is correlated. If this correlation, i.e., the SOKC, is not accounted for, the point estimate of a cut set containing two or more basic events involving failures of these pumps will differ from the true mean value.

To account for this correlation when propagating the basic event values and their uncertainty in a Monte Carlo (or similar) sampling trial, at each pass through the process, the distribution based on the pooled data should be sampled once to obtain a failure rate, and that same failure rate should be used to generate the sample value for all the pump-failure basic events in the cut set equations.

In general then, to account for the SOKC, the same information is used to generate the estimates of the parameters used to evaluate the probabilities of a group of basic events whose parameter values were obtained from correlated data. This means that when using a Monte Carlo (or similar) approach to propagate uncertainty, for each pass through the process the same sample value drawn from the probability distribution of the parameter should be used to calculate the basic event probability of all basic events within the group.

As demonstrated with a simple example in the next section, the effects of the SOKC are that the true propagated mean of the relevant risk metric will have a higher value than the mean value obtained without the correlation, and the parametric uncertainty about the mean will also be underestimated. Therefore, there is a need to understand the significance of this correlation and account for it appropriately. As unique plant-specific data are developed for different component applications, such as motor-operated valves (MOVs) of a certain size range, for example, the impact of the correlation effect will decrease.

6-A.2 Effect of the SOKC on a Risk Metric and its Uncertainty

The mean of a minimal cut set (MCS)[24] containing basic events that are correlated can be underestimated when the SOKC is not accounted for because

$$E(X^n) > E^n(X) \qquad \text{Equation 6-A-1}$$

where X is a random variable corresponding to a basic event that is correlated with other basic events in the MCS, $E(X^n)$ is the expected value of the random variable X elevated to the nth power and $E^n(X)$ is the nth power of the expected value of X.

To illustrate this underestimation for n = 2, consider the simple case where two MOVs are in parallel, represented by variables X_1 and X_2 that are correlated, and system failure occurs when both fail to open. The equation when the failure probabilities of the two MOVs are identical (i.e., the distributions of the failure probabilities express the same state of knowledge) is

$$T = X^2 \qquad \text{Equation 6-A-2}$$

where T represents system failure. This equation expresses the fact that the failure probabilities of the two MOVs are identical (i.e., the distributions of the failure probabilities express the same state of knowledge).

If X_1 and X_2 are considered to be independent, the equation used for system failure would be

$$T = X_1 X_2 \qquad \text{Equation 6-A-3}$$

This equation underestimates the mean of T, as can be seen by taking the expected value in Equations 6-A-2 and 6-A-3. Thus, using Equation 6-A-2, and the definition of variance

$$E(T) = E(X^2) = E^2(X) + \sigma^2_X \qquad \text{Equation 6-A-4}$$

where $E^2(X)$ is the square of the expected value of X and σ^2_X is the variance of X.

Using Equation 6-A-3, with X_1 and X_2 independent, so that their covariance is zero

$$E(T) = E(X_1 X_2) = E^2(X) \qquad \text{Equation 6-A-5}$$

---

[24] An MCS is a minimal set of basic events that causes an undesired outcome, usually in a PRA that outcome is core damage.

Comparing Equations 6-A-4 and 6-A-5 demonstrates that the mean value of the system failure (i.e., the expected value of this failure E(T)) is underestimated when the SOKC is ignored.

The underestimation of the mean of an MCS that contains correlated basic events is particularly significant when

$$E(X^n) >> E^n(X)$$
Equation 6-A-6

This condition occurs when an MCS contains more than two basic events that are correlated or when the uncertainty (i.e., spread) of the distribution of the correlated basic events in an MCS is large.

This example of two MOVs in parallel also serves to illustrate the potential underestimation of the uncertainty as expressed by the variance of the distribution of system failure. (The variance is a measure of spread (i.e., width) of the distribution of the risk metric.) Considering that the distributions of the failure probabilities of the two MOVs express the same state of knowledge, the equation is

$$\sigma^2_T = E(X^4) - E^2(X^2)$$
Equation 6-A-7

where σ^2_T is the variance of T.

If X_1 and X_2 are considered to be independent, the equation used for the variance of the distribution of system failure would be

$$\sigma^2_T = E^2(X^2) - E^4(X)$$
Equation 6-A-8

In typical evaluations, Equation 6-A-7 yields a greater variance (i.e., uncertainty) than Equation 6-A-8. It is important to note that the uncertainty of the distribution of system failure potentially will be underestimated even if the correlated events are not in the same MCS.

Example. The simple system containing two MOVs in parallel is evaluated using data from the paper by Apostolakis and Kaplan [1981] to illustrate the quantitative effect of employing equations for uncorrelated events for estimating the mean and uncertainty expressed in terms of the variance. The data used are the following:

$$E(X) = 1.5 \times 10^{-3}$$
$$E(X^2) = 6.0 \times 10^{-6}$$
$$E(X^4) = 1.6 \times 10^{-9}$$

Using these data, the variance of X is

$$\sigma^2_X = E(X^2) - E^2(X) = 3.8 \times 10^{-6}$$

Table 6-A-1 lists the correlated and uncorrelated values of the mean and variance of system failure and the factor by which these values differ. For this simple example, the mean and variance are underestimated, respectively, by factors of about 2.7 and 50.6.

Table 6-A-1 Example of quantitative effect of failing to account for SOKC.

Parameter	Correlated value	Uncorrelated value	Factor
Mean, $E(T)$	6.0×10^{-6} (Equation 6-A-4)	2.3×10^{-6} (Equation 6-A-5)	2.7
Variance, σ^2_T	1.6×10^{-9} (Equation 6-A-7)	3.1×10^{-11} (Equation 6-A-8)	50.6

Accordingly, failing to take into account the SOKC when evaluating a risk metric or an intermediate value, such as the frequency of an accident sequence, has the following potential impacts:

- For MCSs containing correlated basic events, ignoring the SOKC will underestimate the mean of each MCS containing such events. This point has implications in generating the MCSs from the logic model of the PRA and in using the mean values of the MCSs to estimate the mean of the intermediate results and final risk metrics, as follows:

  - Screening MCSs may delete some of them without justification. This may occur in two contexts. In one, because the number of MCSs in a PRA model can be extremely large, it is common to use a truncation value in solving the PRA's logic model. In this way, only MCSs above this value are obtained, while the rest are neglected. Should the mean value or other point estimate of an MCS be assessed without accounting for the SOKC, and this value used for comparison with the truncation value, some MCSs containing basic events that are correlated may be eliminated when generating the cutsets when they should be retained.

    To illustrate this concern in the first context, assume that a PRA model is evaluated using a truncation value of 1×10^{-9} per year. If the frequencies of the MCSs are calculated using a point estimate that does not account for the SOKC, and the point estimate frequencies of some of the MCSs containing correlated basic events are smaller than this truncation value, a subset of these MCSs may be incorrectly discarded because the correlated frequency (that accounts for the SOKC) of each MCS in this subset is actually larger than this truncation value. The significance of this inappropriate elimination to the estimate of the frequency of a risk metric depends on the combination of two factors: (1) the correct frequency of the risk metric and (2) the correlated frequency of the MCSs in the mentioned subset. If the point estimate frequency of the risk metric is not too large compared to the truncation value, say 1×10^{-6} per year, and the correlated frequency of the MCSs in the mentioned subset is large compared to the truncation value, say 1×10^{-7} per year, the elimination is significant. This issue may become more important for evaluating the mean and the parameter uncertainty in PRAs of future plants, whose risk metrics are expected to have even lower values than those for current plants.

- In evaluating the uncertainty of the risk metric estimate due to propagating the underlying parameter uncertainty, ignoring the SOKC will underestimate the uncertainty.

The combined effect of removing some MCSs (below the truncation value) from the quantitative evaluation and underestimating the mean of MCSs (above the truncation value) containing basic events that are correlated potentially will result in a cumulative underestimation of the mean of the risk metric or other intermediate values.[25]

In summary, failing to take into account the SOKC when evaluating a risk metric or an intermediate value potentially might underestimate the mean and the uncertainty of the distribution of this metric. The simple example above where two MOVs are in parallel clearly reveals that the underestimation of the mean can be significant, especially if the variance of the distribution of the correlated events is large.

---

[25] One approximate approach is proposed in NUREG/CR-4836 for addressing the issues raised in the screening discussion above. The practicality of this approach remains to be demonstrated for large-scale PRAs.

7. STAGE E — ASSESSING MODEL UNCERTAINTY

This section provides guidance to the licensee for addressing sources of model uncertainty and related assumptions[26] related to the base probabilistic risk assessment (PRA) and the application PRA. The goal is to ultimately determine whether (and the degree to which) the risk metric estimates challenge or exceed the quantitative acceptance guidelines due to sources of model uncertainty and related assumptions. Any such source of model uncertainty that could cause the risk metric estimates to challenge or exceed the acceptance guidelines are considered to be key.[27] The process of addressing model uncertainty corresponds to Stage E of the overall process for the treatment of uncertainties.

As discussed in Stage D, a PRA used in a risk-informed application should appropriately account for the uncertainty of the parameters used to quantify the basic events, and this uncertainty should be accounted for in the estimate of the risk metrics by propagating the parameter uncertainty through the accident sequence quantification. This treatment of parameter uncertainty should include, as appropriate, the state-of-knowledge correlation between parameters based on the same data. This section focuses on the treatment of model uncertainty, as defined in Section 2.1.3. Whereas parameter uncertainty is addressed through the methods discussed in Section 6, and becomes part of the base PRA and the PRA modified for the application, the treatment of model uncertainty represents the exercise of sensitivity analyses on the PRA models to assess the potential impact of the choice and selection of models used to construct the base and application PRAs.

7.1 Overview of Stage E

At this point, the base PRA has been modified, as appropriate, to account for (1) any proposed changes to the application (Stages B), and (2) scope limitations relevant to the applications that have been addressed and assessed (Stage C). The modified PRA has been quantified to produce estimates of the appropriate risk metrics for the applications (Stage D). In Stage E the results of the modified PRA are reviewed and analyzed to identify key sources of model uncertainty and related assumptions.

The guidance for addressing the model uncertainty involves two main steps, as illustrated in Figure 7-1.

---

[26] A ***source of model uncertainty*** exists when (1) a credible assumption (decision or judgment) is made regarding the choice of the data, approach, or model used to address an issue because there is no consensus and (2) the choice of alternative data, approaches or models is known to have an impact on the PRA model and results. An impact on the PRA model could include the introduction of a new basic event, changes to basic event probabilities, change in success criteria, or introduction of a new initiating event. A credible assumption is one submitted by relevant experts and which has a sound technical basis. Relevant experts include those individuals with explicit knowledge and experience for the given issue. An example of an assumption related to a source of model uncertainty is battery depletion time. In calculating the depletion time, the analyst may not have any data on the time required to shed loads and thus may assume (based on analyses) that the operator is able to shed certain electrical loads in a specified time.

[27] A source of model uncertainty or the related assumption is considered to be key to a risk-informed decision when it could impact the PRA results that are being used in a decision and, consequently, may influence the decision being made. An impact on the PRA results could include the introduction of a new functional accident sequence, or other changes to the risk profile (e.g., overall core damage frequency (CDF) or large early release frequency (LERF), event importance measures). Key sources of model uncertainty are identified in the context of an application.

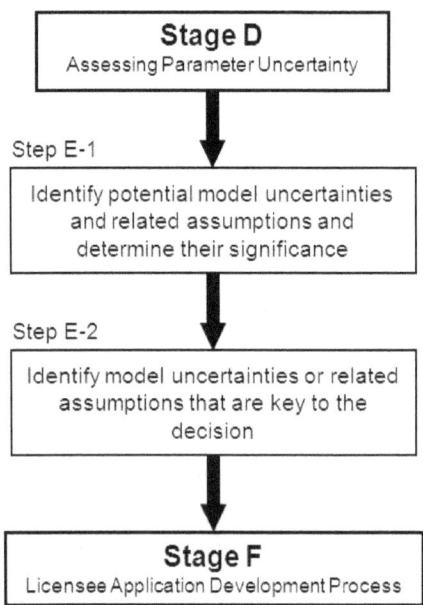

Figure 7-1 Overview of Stage E

- <u>Step E-1</u>: Identify any potential model uncertainties and determine their significance. This step involves identifying sources of model uncertainty in the base PRA. The base PRA is reviewed to identify and characterize the sources of model uncertainty. Some sources may be generic, and some may be plant-specific. These sources of model uncertainty are those that result from developing the PRA. The sources of model uncertainty associated with the base PRA are reviewed to identify those that are relevant to the application under consideration. New sources of model uncertainty that may be introduced by the application also are identified. This identification is based on an understanding of the type of application and the associated acceptance guidelines.

- <u>Step E-2</u>: Identify sources of model uncertainty key to the application. In this step the sources of model uncertainty that are relevant to the application are reviewed to identify those that are key to the application. For the situation where the risk metric calculation from Stage D already challenges or exceeds the acceptance guidelines, it still is necessary to determine the potential significance of these sources of model uncertainty and related assumptions when they result in a further challenge to or exceedance of the acceptance guidelines. This review involves performing a quantitative analysis to identify the importance of each relevant source of model uncertainty.

In providing guidance to the licensee for addressing model uncertainty, this section also provides an acceptable approach for meeting the technical requirements in the ASME and American Nuclear Society (ANS) PRA standard [ASME/ANS 2009]. The relevant requirements in the PRA standard are related to identifying sources of model uncertainty and characterizing the effect of those sources on the PRA (e.g., introduction of a new basic event, changes to basic event probabilities, change in success criterion, or introduction of a new initiating event). As such, the guidance in Section 7.2 for Step E-1 is described in the context of the requirements in the PRA standard. The PRA standard does not provide requirements related to the use of the PRA results in an application; consequently, the guidance provided in Section 7.3 (Steps E-2) is not described in the context of the PRA standard.

Since the guidance in Step E-1 is provided in the context of the PRA standard, it is important to understand the relevant technical requirements (i.e., the supporting requirements (SRs)) in the PRA standard. The relevant SRs that address model uncertainty include the following:

Those that identify sources of model uncertainty and assumptions made in the development of the PRA:

- Part 2—IE-D3, AS-C3, SC-C3,SY-C3, HR-I3, DA-E3, QU-E1, QU-E2, QU-F4, LE-F3,
- Part 3—IFPP-B3, IFSN-B3, IFSO-B3, IFEV-B3, IFQU-A7, IFQU-A10, IFQU-B3,
- Part 4—FQ-E1, FSS-H9, IGN-B5, UNC-A2,
- Part 5—SHA-A1, SHA-D1, SHA-D3, SHA-E2, SHA-F2, SHA-J3, SFR-G3, SPR-B1, SPR-E7, SPR-F3,
- Part 7—WHA-B3, WFR-B3, WPR-A4, WPR-C3,
- Part 8—XFPR-A4, XFHA-B3, XFFR-B3, XFPR-C3,
- Part 9—XHA-A1, XHA-A2, XHA-B3, XFR-A1, XFR-A2, XFR-A4, XFR-B3, XPR-A4, XPR-C3,
- Part 10—SM-H3

Those that identify how the PRA is affected (e.g., introduction of a new basic event, changes to basic event probabilities, change in success criterion, introduction of a new initiating event.):

- Part 2—QU-E4, QU-F4, LE-F3, LE-G4,
- Part 3—IFQU-A7, IFQU-A10,
- Part 4—FQ-E1, FSS-E4, UNC-A2
- Part 5—SHA-C3, SHA-D1, SHA-D3, SHA-E2, SHA-F2, SPR-B1, SPR-E7,
- Part 7—WPR-A4,
- Part 8—XFPR-A4,
- Part 9—XHA-A1, XHA-A2, XPR-A4[28]

7.2 <u>Step E-1</u>: PRA Sources of Model Uncertainty and Related Assumptions—Application Guidelines Met

The purpose of this step is to provide guidance for identifying and characterizing those sources of model uncertainty and related assumptions in the PRA required for the application for the case where the risk metrics calculated in Stage D meet the application acceptance guidelines. Figure 7-2 illustrates the process used to identify and characterize the sources of model uncertainty and related assumptions. This process, discussed in detail in Section 7.2.1 through 7.2.5, involves the following:

- identification of sources of model uncertainty and related assumptions
- characterization of sources of model uncertainty and related assumptions
- qualitative screening of model uncertainty and related assumptions

---

[28] The guidance in this NUREG, along with the guidance in Interim Staff Guidance (ISG) YYY-ISG-X, when followed in the implementation of a base and an application PRA, will ensure compliance with the PRA standard supporting requirements that are relevant to model uncertainty – according to NRC expectations as prescribed in the ISG.

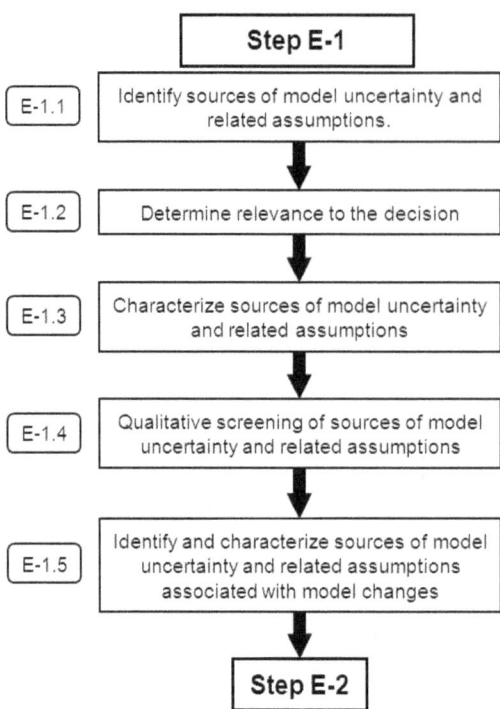

Figure 7-2 Step E-1: Process to identify and characterize the sources of model uncertainty and related assumptions

The process of Step E-1 is consistent with the process discussed in Section 3 of the Electric Power Research Institute (EPRI) report 1016737 [EPRI, 2008] and with the process discussed in Section 4.3 of EPRI 1026511 [EPRI, 2012]. Together, these two EPRI reports provided detailed guidance on the process of identifying and characterizing the sources of model uncertainty and related assumptions.

7.2.1 Substep E-1.1: Identify Sources of Model Uncertainty and Related Assumptions

The purpose of this part of Step E-1 is to identify the sources of model uncertainty that result from developing the PRA. This identification is performed by examining each step of the PRA development process to identify if a model uncertainty or related assumption is involved. This examination provides a systematic method of identifying the sources of model uncertainty. One acceptable process involves using the SRs defined in the PRA standard. Each SR should be reviewed to determine if a model uncertainty or related assumption is involved. Furthermore, this identification process is required by the PRA standard; that is, for each technical element, the PRA standard requires that the sources of model uncertainty be identified and documented.

For this process, EPRI report 1016737 [EPRI, 2008] provides detailed guidance and a generic list of sources of model uncertainty and related assumptions for the internal event hazard group that result from the implementation of the process (see Tables A-1 and A-2 of EPRI report 1016737).

EPRI 1026511 [EPRI, 2012] provides examples of sources of model uncertainty for the internal fires, seismic, Low Power Shutdown and Level 2 hazard groups, in its appendices B, C, D and E. This list can serve as a starting point to identify the set of plant-specific sources of model uncertainty and related assumptions. The analyst is expected to apply the process to also identify any unique plant-specific sources.

For reactors in the preoperational stages (e.g., design certification, combined license application, pre-fuel loading), there may be a lack of design and operational information. Thus, alternative actions may be performed to allow construction of the PRA model. These alternative actions are sources of model uncertainty. Table 7-1 provides a list of activities that cannot be performed during the construction of a design-phase PRA and thus require the analysts to perform alternative actions. Sensitivity analyses may be prudent if these alternative actions prove to be significant to the PRA results. At a minimum, assumptions made in lieu of data, operational experience or design detail should be well documented with the basis for the assumptions clearly explained. An applicant should expect the regulator to hold the applicant to those assumptions as the plant design progresses towards fuel-load. Any changes to such assumptions should be evaluated through updates to the PRA.

Table 7-1 Model limitations that introduce sources of model uncertainty for design-phase PRA.

PRA activity	Issues for design-stage PRA activities	Alternative action and related assumption
Walkdowns	Confirmation of design features using walkdowns will not be possible until construction is completed for the relevant design features.	Drawings (e.g., layout and isometric) and/or 3-D digital simulations can be used as an alternative prior to placement of all equipment. The related assumption is that the plant is configured as designed and passive engineered safety features perform their intended function.
Interviews	Interviews with plant personnel to obtain relevant operating experience will not be possible until plant operation has begun	An alternative action is to interview system designers, procedure developers, and plant personnel from existing plants that have knowledge of the new plant design and proposed operation, as available. The related assumption is that plant personnel will operate the plant as intended by the applicant.

Table 7-1 Model limitations that introduce sources of model uncertainty for design-phase PRA *(Continuation)*.

PRA activity	Issues for design-stage PRA activities	Alternative action and related assumption
Similar Plants	For ALWR and Advanced Non-LWR PRAs for a reference plant design, there may not be similar plants available for comparison.	An alternative action is to evaluate the applicability of data and results from existing plants taking into account differences in design, operational features, environment, and site specific factors, as applicable. The related assumption is that non-nuclear data can be used to help establish a basis for the performance of nuclear structures, systems, and components (SSCs) that share a similar technical pedigree to non-nuclear SSCs with existing data. Furthermore, expert judgment can be used to carefully assess the impact of extenuating circumstances (e.g., high temperatures, new materials).
Plant Specific Operating Experience and Data	Plant-specific operating experience will not be available.	An alternative is to use generic data, and plant specific operating experience from existing plants as applicable. The related assumption is that non-nuclear data can be used to help establish a basis for the performance of nuclear SSCs that share a similar technical pedigree to non-nuclear SSCs with existing data. Furthermore, expert judgment can be used to carefully assess the impact of extenuating circumstances (e.g., high temperatures, new materials).
Treatment of assumptions	Assumptions may be made to develop PRAs for pre-operational plants when design and operational information is preliminary or simply not available.	The uncertainty associated with the assumptions should be identified and the assumptions should be validated when the information becomes available.

At this point in Step E-1.1, a set of model uncertainties and related assumptions has been identified from the base PRA. These sources of model uncertainty and related assumptions have been identified within the context of the SRs of the PRA standard and they have yet to be evaluated for relevancy to the application or impact on the PRA results.

7.2.2 Substep E-1.2: Identify Relevant Sources of Model Uncertainty and Related Assumptions

The purpose of this part of Step E-1 is to identify those sources of model uncertainty and related assumptions in the base PRA that are relevant to an application. Up to this point, Step E-1 has been focused on the base PRA. However, in the context of an application, only part of the base PRA might be relevant and thus, not all sources of model uncertainty from the base PRA may be relevant. The irrelevant sources of model uncertainty and related assumptions may be screened from further consideration.

The process used to identify those sources of model uncertainty from the base PRA relevant to the application involves the following:

- understanding the way in which the PRA is used to support the application (Section 3.3, Step A-3)

- identifying base PRA sources of model uncertainty relevant to the PRA results needed for the application

This understanding and identification involves:

- identifying the results needed to support the application under consideration

- establishing the changes to the PRA that are required to accurately reflect the proposed plant changes in the application

Screening Based on Relevance to the Needed Parts of the PRA

Some sources of model uncertainty may be screened because they are only relevant to parts of the PRA that are not exercised by the application. For example, if the application is only concerned with the loss-of-coolant accident (LOCA) sequences, then those sources of model uncertainty and related assumptions impacting the other sequences would not be considered relevant to the application. Only those sources of model uncertainty that affect the parts of the base PRA needed to support the application need to be retained for further evaluation. For example, when the application addresses an allowed outage time extension for a diesel generator, only those parts of the PRA that involve the diesel generator need be exercised, namely those sequences that involve a loss of offsite power (LOOP). Thus, only uncertainties that affect these LOOP sequences would have to be considered. However, it should be noted that for an application such as this that uses the acceptance guidelines in Regulatory Guide (RG) 1.174, Revision 1, "An Approach for Using probabilistic Risk Assessment in Risk-Informed Decisions on Plant-Specific Changes to the Licensing Basis," issued November 2002 [NRC, 2002], the total core damage frequency (CDF) also may need to be determined depending on the value of the change in CDF (ΔCDF). In these cases, the uncertainties in the complete base PRA would need to be retained for further consideration.

On the other hand, an application such as the implementation of 10 CFR 50.69, "Risk-Informed Categorization and Treatment of Structures, Systems, and Components for Nuclear Power Reactors," requires the categorization of SSCs into low- and high-safety significance. Because these are relative measures, such an assessment would involve the complete PRA. RG 1.201, "Guidelines for Categorizing Structures, Systems, and Components in Nuclear Power Plants According to Their Safety Significance," [NRC, 2006a] and the Nuclear Energy Institute (NEI) Report 00-04, "10 CFR 50.69 SSC Categorization Guideline," issued April 2004 [NEI, 2005b], address treatment of uncertainties in the categorization of SSCs. The guidelines provided in this NUREG do not supersede the approach in 10 CFR 50.69.

7.2.3 Substep E-1.3: Characterization of Sources of Model Uncertainty and Related Assumptions

The purpose of this part of Step E-1 is to characterize the identified sources of model uncertainty and related assumptions. This characterization involves understanding how the identified sources of model uncertainty and related assumptions can affect the PRA. The following must be identified (as required by QU-E4 of the PRA standard) when characterizing sources of model uncertainty and related assumptions:

- the part of the PRA affected

- the modeling approach or assumption used

- the impact on the PRA (e.g., introduction of a new basic event, changes to basic event probabilities, change in success criteria, introduction of a new initiating event)

- identification of conservative bias

7.2.3.1 Identifying the Part of the PRA Affected

In this part of the characterization process, the part of the PRA affected by the source of model uncertainty or a related assumption is identified. This identification is needed because not every application involves every aspect of the PRA. Therefore, as discussed below in Step E-2, if the application deals with an aspect of the PRA not affected by the source of model uncertainty, then the source is not relevant to the application.

The sources of model uncertainty could impact the PRA by affecting the following:

- a single basic event
- multiple basic events
- the logic structure of the PRA event trees or fault trees
- a combination of both basic events and portions of the logic structure

Model uncertainties and related assumptions exist that can influence the frequency of initiating events, human error probabilities (HEPs), the failure probabilities of structures, systems, and components (SSCs), or the PRA fault tree and event tree models. For example, different approaches are available to generate LOCA initiating event frequencies. Uncertainty in deterministic analyses can also influence basic event failure rates. For example, the choice of assumptions used in deterministic calculations used to assess sump plugging will influence the probability assigned to that event. An uncertainty associated with the establishment of the success criterion for a system could be whether one or two pumps are required for a particular scenario. This uncertainty would be reflected in the choice of the top gate in the fault tree for that system.

7.2.3.2 Modeling Approach Used or Assumption Made

In this part of Step E-1.3, the potential impact of each identified source of model uncertainty on the PRA models and results must be determined. Determining a source's potential impact involves identifying how the PRA results would change if an alternate model were selected. The following is a list of some examples of how an alternate model could impact the PRA:

- an alternate computational model may produce different initiating event frequencies, SSC failure probabilities, or unavailabilities

- an alternate human reliability analysis (HRA) model may produce different HEPs or introduce new human failure events

- an alternate assumption regarding phenomenological effects on SSC performance can impact the credit taken for SSCs for some accident sequences

- an alternate success criterion may lead to a redefinition of system fault tree logic

- an alternate screening criterion may lead to adding or deleting initiating events, accident sequences, or basic events

- an alternate assumption that changes the credited front-line or support systems included in the model may change the incremental significance of sequences

- an alternate seal LOCA model can result in a different event tree structure, different seal LOCA sizes, and different seal probabilities

7.2.3.3 Identification of Conservatism Bias

An important aspect of characterizing a source of model uncertainty or related assumption is to understand whether the chosen model or adopted assumption is conservative. This understanding is necessary because for some applications, the use of conservative assumptions in one part of the model can mask the significance of another part of the model—a part of the model that might be needed for the application. This is particularly true for applications that involve risk categorization or risk ranking. Thus, if a source of model uncertainty or related assumption is identified as resulting in a conservative bias, the impact of this bias on the conservatism in the PRA must be assessed.

A conservative bias, as used in this report, implies that adopting a conservative model or assumption would lead to a higher risk estimate than if a more realistic model or assumption was adopted. A de-facto consensus of acceptance may exist when certain conservative NRC licensing criteria are used as the basis to model certain issues. An example of such a conservative criterion is the 2- to 4-hour coping time for battery depletion during an event involving loss of alternating current (ac) power. This coping time would be considered conservative because station batteries are expected to be available for several more hours if loads are successfully shed. A model that reflects a plant's licensing basis is generally perceived to have a conservative bias because it incorporates the conservative attributes of the deterministic licensing criteria.

At this point in Step E-1, a set of model uncertainties and related assumptions has been identified as relevant to the application (Steps E-1.1 and 1.2) and characterized to assess their impact on the PRA (e.g., introduction of a new basic event, changes to basic event probabilities, change in success criteria, introduction of a new initiating event), and in the identification of conservative bias.

7.2.4 Substep E-1.4: Qualitative Screening

The purpose of this part of Step E-1 is to qualitatively screen out sources of model uncertainties. This screening identifies those sources of model uncertainty that do not warrant further consideration as potential key sources of model uncertainty and related assumptions. That is, sources of model uncertainty or related assumptions may exist that, for qualitative reasons, do not need to be considered as potentially key to the application. This qualitative screening involves identifying and validating whether consensus models have been used in the PRA to evaluate identified model uncertainties.

Consensus Model

The use of a consensus model eliminates the need to explore an alternative hypothesis, but adoption of a consensus model does not mean that the consensus model has no uncertainty associated with its use. However, this uncertainty would generally be manifested as an uncertainty on the parameter value or values used to generate the probability of the basic event(s) to which the consensus model is applied. This uncertainty would be treated in the PRA quantification as a parameter uncertainty. The adoption of a consensus model obviates the need to consider other models as alternatives.

There may be cases where there may be more than one consensus model for addressing a specific issue. An example is the Multiple Greek Letter and the Alpha methods for quantifying common cause failures. In such a case, any one of the consensus models can be used. Multiple consensus models should provide similar results. If they do not, then they do not meet the requirement for being a consensus model and an evaluation of the associated model uncertainty should be made.

The models used in the PRA are reviewed to identify those that meet the definition of a consensus model and, consequently, can be screened from further consideration. The definition of a consensus model (given in Section 2.1.3) is as follows:

> **Consensus model -** In the most general sense, a consensus model is a model that has a publicly available published basis[29] and has been peer reviewed and widely adopted by an appropriate stakeholder group. In addition, widely accepted PRA practices may be regarded as consensus models. Examples of the latter include the use of the constant probability of failure on demand model for standby components and the Poisson model for initiating events. For risk-informed regulatory decisions, the consensus model approach is one that the NRC has utilized or accepted for the specific risk-informed application for which it is proposed.

It is important to note that, by this definition, a legitimate consensus model is characterized by both the model itself and on how it is used within the context of a specific application. This relationship exists because consensus models have limitations that may be acceptable for some uses and not for others. Some examples of consensus models include the following:

- Poisson model for initiating events
- Bayesian analysis
- Westinghouse and Combustion Engineering reactor coolant pump seal LOCA models
- Multiple Greek Letter and Alpha method for evaluating common cause failures

Currently there is no agreed-on list of consensus models nor is there a formal process to establish such a list. However, as a first step in establishing such a process, EPRI has compiled a list of candidate consensus models [EPRI, 2006a]. This list includes common approaches, models, and sources of data used in PRAs. At this time, the NRC has not reviewed this list although specific models, approaches and data may have been approved for specific risk-informed applications.

---

[29] It is anticipated that most consensus models would be available in the open literature. However, under the requirements of 10 CFR 2.390, there may be a compelling reason, for exempting a consensus model from public disclosure.

At this point in Step E-1a set of candidate model uncertainties and related assumptions have been identified, characterized, and some of these have been screened out from further consideration because of the use of consensus models or because they have been deemed irrelevant to the application.

7.2.5 Substep E-1.5: Identifying and Characterizing Relevant Sources of model uncertainty Associated with Model Changes—Modified PRA.

The purpose of this part of Step E-1 is to identify any new sources of model uncertainty that may be introduced as a result of the application. Modifications are made to the PRA to represent the effect of a potential change or to investigate a different design option. The modifications themselves may introduce new sources of model uncertainty and related assumptions, which need to be identified. The process used in Section 7.2.1 through 7.2.4 to identify sources of model uncertainty in the base PRA is repeated for the modifications to the PRA. Specifically, the PRA modifications are reviewed against the applicable PRA standard SRs to determine if a model uncertainty or related assumption has been introduced. For example, to assess the impact of relaxing the special treatment requirements for SSCs with low-risk significance, it is necessary to model the impact on the SSC reliability. No accepted model exists for this impact; therefore a model uncertainty has been introduced.

> Tables A-3 and A-4 in EPRI report 1016737 [EPRI, 2008] provide additional information that could be useful in searching for sources of model uncertainty and related assumptions that typically are not significant contributors to the base PRA, but have been noted as significant in specific applications. This helps to ensure that the PRA is comprehensively evaluated for relevance to the application.

At this point in Step E-1, a set of candidate model uncertainties and related assumptions has been identified and characterized, and some of these have been screened out from further consideration because of the use of consensus models or because they have been deemed irrelevant to the application. This list of candidate model uncertainties includes issues associated with both the base PRA and the modified PRA.

7.3 Step E-2: Identification of Key Sources of Model Uncertainty and Related Assumptions—Application Acceptance Guidelines Met

The purpose of this step is to provide guidance for identifying those sources of model uncertainty and related assumptions that are key to the application when the application has met the application acceptance guidelines in Stage D (Section 6). This step involves identifying those sources of model uncertainty and related assumptions key to the application. Although a source of model uncertainty may be relevant, its actual impact on the results may not be significant enough to challenge the application's acceptance guidelines. Only the relevant sources of uncertainties and related assumptions with the potential to challenge the application's acceptance guidelines are considered key.

In those situations where the calculation of the risk metric in Stage D (Section 6) has already challenged or exceeded the acceptance guidelines, it still is necessary to determine the potential significance of these sources of model uncertainty and related assumptions to further challenge or exceed the acceptance guidelines (e.g., moving further into Region I in Figure 7-4).

When this is the case, any source of model uncertainty and related assumption that results in a further challenge or exceedance of the acceptance guidelines should be assessed in the license application as if they are key model uncertainties.

The input to this step is a list of the relevant sources of model uncertainty identified in Step E-1. These sources of model uncertainty and related assumptions may now be quantitatively assessed to identify those with the potential to impact the results of the PRA such that the application's acceptance guidelines are challenged or are further challenged or exceeded, as in the case where the Stage D risk metric calculation already challenges the acceptance guidelines. This assessment is made by performing sensitivity analyses to determine the importance of the source of model uncertainty or related assumption to the acceptance criteria or guidelines. Those determined to be important are key sources of model uncertainty and related assumptions. Figure 7-3 illustrates the general process.

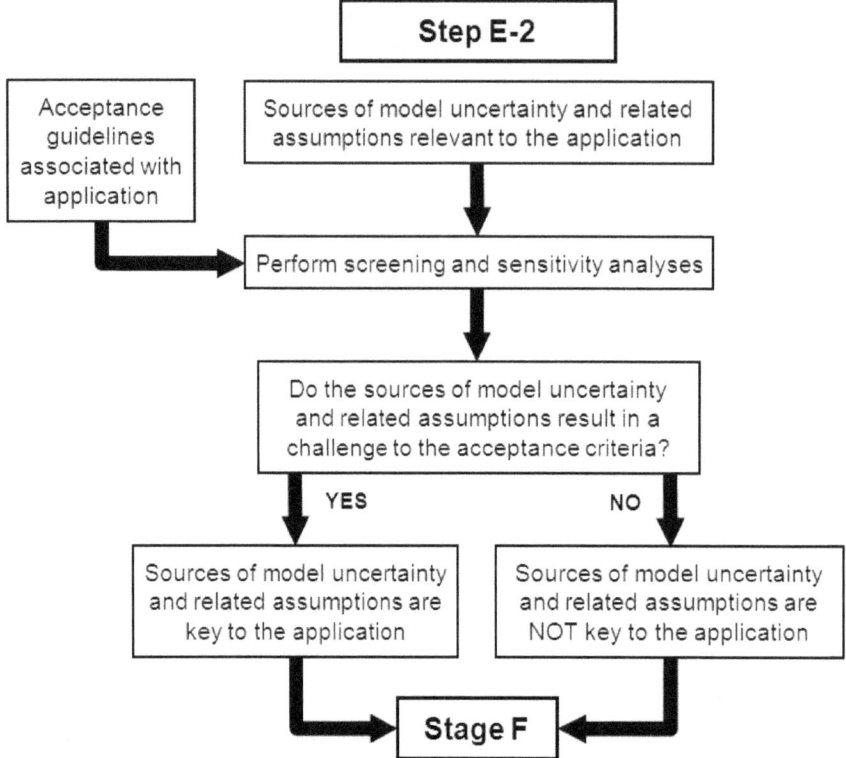

Figure 7-3 Step E-2: Process to identify key sources of model uncertainty and related assumptions

The process for identifying the key sources is, in principle, straightforward; however, it is dependent on the nature of the uncertainty being assessed and on the nature of the acceptance guidelines, which adds some complexity. The process involves the following:

- defining and justifying the sensitivity analyses
- performing the sensitivity analyses

The sensitivity analysis determines whether the application's acceptance guidelines are challenged because of the model uncertainty. The acceptance guidelines may consist of either a single metric or multiple metrics. For each type of acceptance guideline (one or two metrics), a sensitivity analysis is performed to screen out those that are not key to the application. Section 7.3.2 discusses the single metric case, referred to as Case 1. Section 7.3.3 discusses the two-metric case, referred to as Case 2.

Moreover, for Case 1 and Case 2 the following different options can be employed in performing the sensitivity analysis:

- Conservative Screening—Perform a conservative screening sensitivity analysis. If a model uncertainty can be shown not to be key, as based on a conservative model, then additional realistic sensitivity analyses are not needed. If a conservative sensitivity analysis could result in a key model uncertainty, then it is necessary to continue the analysis with realistic sensitivity analyses.

- Realistic Sensitivity—Perform a realistic sensitivity analysis. This may be performed if the analyst does not consider a conservative screening sensitivity analysis to be useful or practical.

A realistic analysis involves developing credible alternatives or hypotheses associated with the model uncertainties relevant to the application. An alternative or hypothesis is considered to be credible when it has a sound technical basis (see footnote 22). The set of sensitivity analyses needed to obtain a credible understanding of the impact of the source of model uncertainty or related assumptions is dependent on the particular model uncertainty or related assumption. To develop the alternatives, an in-depth understanding of the issues associated with the model uncertainty or related assumption is needed. What is known about the issue itself will likely dictate the possible alternatives to be explored.

One example of developing alternative models is the use of previous experience in PRAs. Variations in the way a particular model uncertainty or related assumption has been addressed in other PRA analyses, both for base PRA evaluations and for related sensitivity analyses could provide credible alternatives. In general, these variations in addressing a model uncertainty will have been accepted as credible in the literature. An example is the operating equipment reliability—following a loss of room cooling—for which no specific calculations exist. An accepted conservative model assumption is to assign a failure probability of 1 for the equipment in question. Alternatively, it may be worthwhile to explore whether the issue of room cooling is relevant to the application by performing a sensitivity analysis under the assumption that room cooling is not needed for successful equipment operation.

Another example for developing alternative models involves varying a parameter value for the purpose of deriving a range of credible values for the parameter. For example, consider the issue of battery life. If a conservative licensing-basis model has been used, consider increasing battery life by 50 percent to represent the potential to extend battery life through operator actions. Alternatively, if a credible life extension model has been used that allows for load shedding, such a model assumes that the load-shedding procedures are performed successfully. To test the potential uncertainty impact of this assumption, one can quantify the PRA model with a 50 percent reduction in battery life to reflect the possibility that equipment operators fail to successfully perform all tasks under stressful conditions.

For other sensitivity analyses, it is common to change the value of specific parameters or groups of parameters by factors of 2, 5, or 10. For these sensitivity analyses to be meaningful, it is necessary to have a justification for the factors based on an understanding of the issue responsible for the uncertainty. An alternative approach to justifying the factor is to implement a performance monitoring program that would verify that the assumed factor is justified. Section 8.4 discusses this topic as part of developing a strategy to address key model uncertainties.

Section 4.3.1 of EPRI report 1016737 [EPRI, 2008] and Section 4.4.1 of EPRI 1026511 [EPRI, 2012] provide guidance on determining a reasonable range over which a sensitivity analysis should investigate model uncertainty.

7.3.1 Case 1: Applications Involving a Single-Metric Acceptance Guideline

The purpose of the second part of Step E-2 is to provide guidance for performing conservative screening or a realistic assessment for applications involving a single-metric acceptance guideline. In performing the sensitivity analysis, the sources of model uncertainty and related assumptions are linked to the four main ways in which model uncertainty can impact a PRA, as discussed in Step E-1.3 (Section 7.2.3.1):

- a single basic event (Case 1a)
- multiple basic events (Case 1b)
- the logic structure of the PRA (Case 1c)
- logical combinations (Case 1d)

For each case, guidance is provided for both a conservative option and a realistic option.

The concept of "acceptable change in risk" needs to be defined within the context of the application for which the licensee intends to use the PRA. However, it would most likely be defined in terms of a maximum acceptable value for a risk metric, such as CDF, incremental CDF deficit, or incremental core damage probability.

7.3.1.1 Case 1a: Sources of Model Uncertainty and Related Assumptions Linked to a Single Basic Event

For Case 1a, the sources of model uncertainty and related assumptions identified in Step E-1 are reviewed to determine those that are relevant only to a single basic event. For each identified source of uncertainty, the sensitivity of the PRA's results to alternative hypotheses is assessed. Two approaches can be used. The first is a conservative screening option that uses methods such as risk reduction and risk achievement worth (RAW) importance measures [Modarres, 2006]. The second approach is to use realistic sensitivity assessments, in which alternative hypotheses are used that are based on a realistic assessment of data, operational experience, or analysis and research.

Conservative Screening Option

An approach to determining the importance of a source of model uncertainty and related assumptions is to calculate a maximum acceptable RAW, denoted here as RAW_{max}, associated with the risk metric of interest, such as maximum allowable CDF or LERF. The RAW for each relevant basic event can be compared to the RAW_{max} associated with the maximum acceptable CDF. For a basic event with a RAW less than RAW_{max}, the model uncertainty or related assumption associated with that basic event is not, by itself, considered key since it is mathematically impossible for the risk metric to exceed the maximum acceptable value, even when it is assumed that the basic event is certain to occur (i.e., has a failure probability equal to 1).

For the jth basic event, the definition of RAW is

$$RAW_{j,base} = \frac{CDF^+_{j,base}}{CDF_{base}}$$

Equation 7-1

where

RAW_j is the value of RAW for basic event j as calculated in the base PRA,

CDF_{base} is the value of the CDF mean estimate in the base PRA,

$CDF^+_{j, base}$ is the base PRA CDF mean estimate with the basic event j set to 1.

Thus, given that the acceptance criterion defines the maximum acceptable CDF (shown here as CDF^+), Equation 7-1 can then be solved directly for the maximum acceptable RAW (shown here as RAW_{max},) that any basic event can have before the associated model uncertainty or related assumption is considered as being a key uncertainty. To determine RAW_{max}, CDF^+ is substituted into Equation 7-1 in place of $CDF^+_{j, base}$, which changes $RAW_{j, base}$ into RAW_{max}, and the equation is solved for RAW_{max} as shown in Equation 7-2.

$$RAW_{max} = \frac{CDF^+}{CDF_{base}}$$

Equation 7-2

To illustrate the concept of a maximum acceptable RAW, suppose that CDF_{base} for a particular base PRA is 3.0×10^{-5}/year (yr). Suppose further that the maximum acceptable CDF for a particular application of the base PRA, CDF^+, is 5.0×10^{-5}/yr. Hence, from Equation 7-2

$$RAW_{max} = \frac{5.0 \times 10^{-5}}{3.0 \times 10^{-5}}$$

$$RAW_{max} = 1.7$$

For this example, consider a basic event from the base PRA that has a RAW greater than 1.7 and is associated with some source of model uncertainty or related assumption. That is, if this basic event were guaranteed to occur (i.e., the failure probability is equal to 1) the calculated CDF would exceed the maximum acceptable CDF for that application. Based on the RAW for that basic event, the associated source of model uncertainty or related assumption is considered to be potentially key. If a basic event has a RAW greater than the maximum acceptable RAW, the source of model uncertainty or related assumption associated with that basic event is potentially key and cannot be excluded from the analysis. The general expression for any basic event is expressed in Equation 7-3.

$$RAW_{j,base} > RAW_{max}$$

Equation 7-3

If the expression in Equation 7-3 is true for the jth basic event, then the model uncertainty or related assumption associated with the jth basic event is potentially key to the application. The model uncertainty or related assumption should then be assessed with a credible and realistic sensitivity analysis to determine whether it truly is a key source of model uncertainty or related assumption. If Equation 7-3 is false (i.e. $RAW_{j, base} \leq RAW_{max}$), the model uncertainty or related assumption does not require further consideration.

The result in Equation 7-3 is relevant only for those sources of model uncertainty, as identified in Step E-1, that are associated with a single basic event.

Realistic Sensitivity Assessment Option

A realistic sensitivity assessment would determine the basic event probability that would cause the expression in Equation 7-3 to change from false to true. If this so-called "transitional" basic event probability uses an alternative hypothesis based on a realistic assessment of data, operational experience or analysis and research, then the source of model uncertainty or related assumption is key to the application. However, in some cases where it might be difficult to posit a credible hypothesis, one might demonstrate that the value needed to achieve the transitional basic event probability is unreasonably high—thereby making the argument that any credible model would not result in a transitional value.

7.3.1.2 Case 1b: Sources of Model Uncertainty and Related Assumptions Linked to Multiple Basic Events

For Case 1b, the sources of model uncertainty and related assumptions identified in Step E-1 are reviewed to identify those that are relevant only to multiple basic events. An example is an assumption that affects the quantification of a particular failure mode of several redundant components (e.g., stuck-open safety relief valve). Similar to the case of a single basic event, a conservative or realistic analysis may be applied. For each identified source of uncertainty, the sensitivity of the PRA's results to alternative hypotheses is assessed. Two approaches can be used. The first is a conservative screening option which uses methods such as Risk Reduction and Risk Achievement Worth importance measures. The second approach is to use realistic sensitivity assessments, in which alternative hypotheses based on a realistic assessment of data, operational experience or analysis and research is used.

Conservative Screening Option

The RAW importance measures for a given group of basic events, which share a common source of model uncertainty or related assumption, cannot be used to collectively assess the combined impact on the application of that source of model uncertainty (i.e., adding together the individual RAW importance measures of the basic events in a group does not result in a "group" RAW importance measure). Thus, to determine the bounding impact on the risk model of the model uncertainty or related assumption associated with that group of events, the basic event probability for each event in that group must all be set to a value of 1 and the entire PRA model must be requantified.[30] Similar to Equation 7-3, Equation 7-4 establishes the criterion used to determine whether the source of model uncertainty or related assumption is key to the application or if it may be screened out of the application.

$$CDF^+_{k,base} > CDF^+ \qquad \text{Equation 7-4}$$

---

[30] The option of setting a basic event value to logically **TRUE** rather than 1, available in many PRA software packages, is not advised in the context of performing the conservative screening analysis. Doing so risks losing information when the PRA is solved because the use of **TRUE** eliminates entire branches of fault trees and, hence, results in the loss of cut sets from the final solution.

In this case, the subscript "k" represents the kth group of basic events that have a common source of model uncertainty or related assumption. Thus, $CDF^+_{k,base}$ represents the CDF that results from setting equal to 1 the basic event probabilities of the kth group in the base PRA model. Similar to Equation 7-2, in this equation, CDF^+ represents the maximum acceptable CDF for the application. If the expression in Equation 7-4 is true, then the source of model uncertainty or related assumption is potentially key to the application and should be evaluated with a realistic sensitivity analysis. If the expression in Equation 7-4 is false, then, as long as it is not a member of a logical combination (see Section 7.3.1.4), the source of model uncertainty or related assumption is not key to the application. It would be mathematically impossible that any quantification of the basic events associated with that source of model uncertainty or related assumption could produce an unacceptably high CDF.

Realistic Sensitivity Assessment Option

Similar to the case of a single basic event, the realistic sensitivity assessment for multiple events should employ alternative models or assumptions based on realistic assessments of data, operational experience or analysis and research for the group of basic events in order to produce new alternate basic event probabilities for each basic event. For each credible alternative model, the base PRA should be requantified using the alternate basic event values. The result is a set of new CDF values, which represent the range of potential PRA results, each of which needs to be compared against the acceptance criterion dictated by Equation 7-4. If any one of the requantification results exceeds or challenges the application's acceptance guidelines, then the source of model uncertainty or related assumption is considered key to the application.

7.3.1.3 Case 1c: Sources of Model Uncertainty and Related Assumptions Linked to the Logic Structure of the PRA

For Case 1c, the sources of model uncertainty and related assumptions identified in Step E-1 are reviewed to determine those that are relevant only to the logic structure of the PRA.

Alternative methods or assumptions need to be assessed by manipulating or altering the PRA to reflect these alternatives. The methods and assumptions in question are those that could possibly introduce:

- new cut sets in existing sequences by changing fault tree models
- new sequences by changing the structure of event trees
- entirely new classes of accident sequences by introducing new initiating events

Credible uncertainty analyses for sources of model uncertainty and related assumptions should be conducted for each affected logic structure to evaluate whether the sources of model uncertainty and related assumptions are key to the application.

Conservative Screening Option

The effort to change the PRA logic structure and requantify the accident sequences can involve significant resources. However, in some cases, it may be possible to perform an approximate bounding evaluation (see Section 5.2, Step C-1) that demonstrates that the potential impact of the alternate assumption or model would not result in a risk metric that challenges the acceptance guidelines. For example, this demonstration can be achieved if the model uncertainty or related assumption is relevant only to the later event tree branch points of relatively low frequency accident sequences. If, in these accident sequences, the frequencies of the partial accident sequences modeled by the portion of the event tree before those relevant late branch points are sufficiently low to ensure that the acceptance guidelines would be met—regardless of the potential impact of the uncertainty on the remaining accident sequence model—then the model uncertainty would not be key to the application.

Realistic Sensitivity Assessment Option

The analyst should select credible alternative models or assumptions for the particular source of model uncertainty or related assumption and make the required changes to the PRA logic model. Then, for each credible alternate model or assumption, the analyst should requantify the base PRA and reevaluate, using Equation 7-4, the relationship between the maximum acceptable CDF (CDF^+) and the new CDF estimate. These new CDF values constitute a set of potential base PRA results (one for each hypothesis tested). If any one of the requantification results exceeds or challenges the application's acceptance guidelines, then the source of model uncertainty or related assumption is considered key to the application.

7.3.1.4 Case 1d: Sources of Model Uncertainty and Related Assumptions Linked to Logical Combinations

The sources of model uncertainty or related assumptions identified in Step E-1 are reviewed to determine those that are relevant to combinations of basic events and logic structure. In this context, the combinations of basic events and logic structure are termed "logical combinations" and the associated sources of model uncertainty are known as a "logical group of sources and assumptions." Guidance is provided for determining the logical combinations and for performing conservative and realistic sensitivity analyses.

Logical Combinations

For these cases, the logical combination may impose a synergistic impact on the uncertainty of the PRA results. The resulting uncertainty from their total impact may be greater than the sum of their individual impacts. For example, several sources of model uncertainty could relate to the same dominant cut sets, certain sequences, a particular event and the success criteria for systems on that event tree, or to the same plant damage state. In other words, such sources of model uncertainty overlap by jointly impacting the same parts of the risk profile modeled in the PRA. Thus, to accurately assess the full potential for the impact of uncertainty, such sources of model uncertainty also should be grouped together.

An example of this type of impact is demonstrated by examining the relationship of two models—recovery of offsite power and recovery of failed diesel generators—to the overall uncertainty of the PRA model. Both models represent the failure to restore ac power to critical plant systems through different but redundant power sources. Hence, the potential total impact of uncertainty associated with the function of supplying ac power to emergency electrical buses would involve a joint assessment of the uncertainty associated with both models. Another example is the interaction between uncertainties associated with the direct current (dc) battery depletion model and those models associated with operator actions to restore power; specifically, the interrelationship between operator performance and the performance of key electrical equipment under harsh conditions (e.g., smoke, loss of room cooling). The length of time that dc batteries can remain sufficiently charged and successfully deliver dc power to critical components depends on the shedding of nonessential electrical loads, which is achieved through the actions of reactor operators and equipment operators as well as through procedures and the availability of required tools (e.g., lighting, procedures, and communication devices). The uncertainty associated with these operator actions and the impacts of potential harsh environmental conditions on both operators and equipment should be jointly assessed to gain a more holistic understanding of the potential total impact of uncertainty on the dc battery depletion model.

Moreover, the choice of HRA method can impact the uncertainty of PRA results in several areas. An interface exists between the human actions necessary to restore diesel generator operation after either failing to start or run and the time until dc battery depletion occurs. Many diesel generators depend on dc power for field flashing for a successful startup. If equipment operators fail to successfully restore diesel generator operation before the dc batteries are depleted, then the diesel generators cannot be restored to operation. Hence, the potential impact of uncertainties associated with the HEPs in the diesel generator recovery model and uncertainties associated with the dc battery depletion model should be assessed together.

In the above examples, the uncertainty issues are linked by their relationship to a given function, namely, establishing power. However, uncertainties related to different issues can also have a synergistic effect. For example, consider the uncertainty associated with the modeling of a high-pressure coolant injection (HPCI) system in a PRA for a boiling-water reactor. In core damage sequences associated with event trees for transient initiating events, failure of the HPCI is either coupled with failure of other high-pressure injection systems (reactor core isolation cooling, recovery of feedwater, control rod drive) and failure of depressurization, or failure of other high-pressure injection systems and failure of low-pressure injection systems. The importance of HPCI is affected by the benefit realized for the successful operation of additional injection systems. For example, realizing the benefit of fire water injection (as an additional low-pressure system), control rod drive, or recovery of feedwater (as a high-pressure system) can reduce the importance of HPCI.

In the LOOP/station blackout (SBO) event tree, a significant function of HPCI is to give time to recover the offsite power. Therefore the importance of HPCI is affected by the modeling of recovery of offsite power in the short term (assuming that HPCI has failed), the frequency of LOOP events, and the common cause failure (CCF) probability of the diesels and the station batteries.

In summary, the importance of the HPCI system is affected by the following:

- frequency of transient initiating events

- HEP for depressurization

- credit for motor-driven feedwater pumps

- credit for alternate injection systems (e.g., fire water, service water cross-tie)

- LOOP initiating event frequency, common-cause failure of diesels and batteries, and the factors associated with the short-term recovery of ac power given a LOOP.

In general, uncertainties associated with any of these issues could interact synergistically to impact the overall model uncertainty associated with the modeling of the HPCI.

Once the various logical combinations that give rise to sources of model uncertainty have been identified, the quantitative assessment can be performed.

> Further guidance on grouping issues into logical groupings can be found in Section 4.3.2 of EPRI report 1016737 [EPRI, 2008]. The analyst's judgment and insight regarding the PRA should yield logical groupings specific to the PRA in question. Certain issues may readily fall into more than one logical grouping depending on the nature of the other issues.

Conservative Screening Option

When all the contributors to a logical combination of sources and assumptions impact only basic events, the conservative screening approach is similar to Case 1b with regard to quantitative screening. That is, to determine the true impact on the risk model from the logical group of sources associated with a given logical combination, all basic event probabilities associated with that logical combination should be set to 1, and the PRA model should be requantified. Guidance for performing a conservative screening is exactly the same as for Case 1b, discussed in Section 7.3.1.2.

If the logical combinations involve impacts on both the basic event values and the PRA structure, the process of performing conservative screening becomes more involved. The impacts on the PRA structure should be evaluated first so that the impacts of on the basic event probability values can then be assessed with the modified logic structure. The approach described in Case 1c (Section 7.3.1.3) can address the effects of the sources of model uncertainty on the PRA logic. The modified PRA structure can then be used in conjunction with the process in Case 1b (Section 7.3.1.2) to assess the effect of a credible alternative model on multiple basic events, as evaluated using the modified PRA structure. Similarly, Equation 7-4 is used to determine whether the impacts exceed the application acceptance guideline.

Realistic Sensitivity Assessment Option

Similar to previous realistic sensitivity assessment options, credible alternative models should be selected for a given logical group of sources and assumptions and the PRA logic and basic event probability values should be modified based on the selected alternative model; the logic is modified first and then the basic event probabilities. For each credible alternative model, the PRA is requantified. The new risk metric values constitute a range of potential base PRA results. If any one of the requantification results exceeds or challenges the application's acceptance guidelines, then the logical group of sources and assumptions is considered to be potentially key to the application.

7.3.2 Case 2: Applications Involving a Two-Metric Acceptance Guideline

The purpose of this section is to provide guidance for performing screening or assessment for each of the four classes of screening analysis for applications involving a two-metric acceptance guideline. In general, these types of applications are license amendment applications. For example, quantitative assessment of the risk impact in terms of changes to the CDF (i.e., ΔCDF) (or LERF)[31] metric is compared against the RG 1.174 acceptance guidelines (Figure 7-4) or guidelines derived from those of RG 1.174.

---

[31] The discussion that follows is in terms of CDF but can also be applied to LERF.

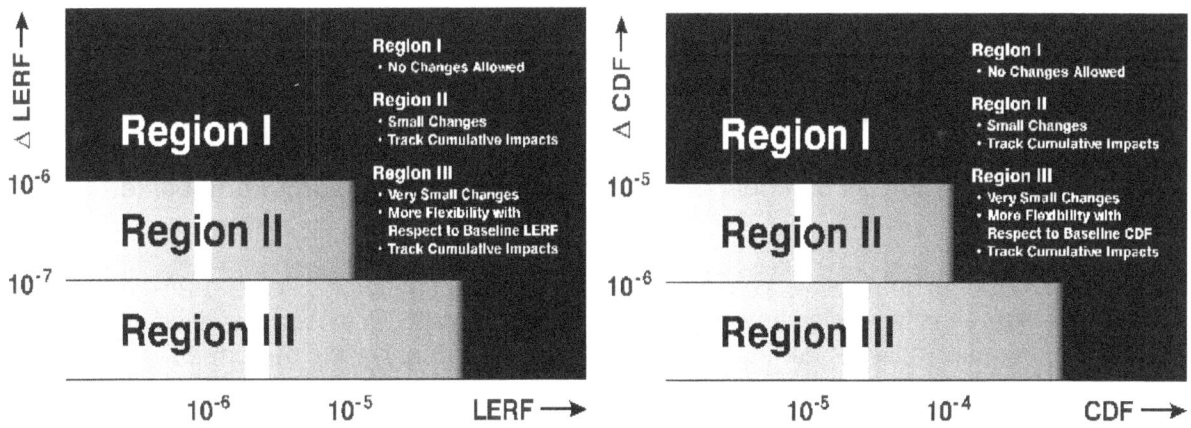

Figure 7-4 NRC Regulatory Guide 1.174 acceptance guidelines for CDF and LERF

Because the acceptance guidelines involve two metrics (CDF on the horizontal axis, and ΔCDF on the vertical axis) it is necessary to assess the potential impact of a model uncertainty issue with respect to CDF and ΔCDF since the acceptability result is based on the relative position within the figure. Hence, just as for applications involving only one risk metric(see Section 7.3.1), it is necessary to assess the potential impact of model uncertainty on the CDF of the base PRA; however, it is also necessary to assess the potential impact of model uncertainty on the ΔCDF. Therefore, the following metrics are of interest for applications involving a change to the licensing basis:

CDF_{base} the value of the CDF mean estimate in the base PRA

CDF_{app} the value of the CDF mean estimate in the modified base PRA to account for changes proposed to the licensing basis

$CDF^{+}_{j,\,base}$ the CDF mean estimate in the base PRA with the basic event j set to 1

$CDF^{+}_{j,\,app}$ the CDF mean estimate in the modified PRA with the basic event j set to 1

Using these four quantities, the terms ΔCDF and ΔCDF⁺ are defined as follows:

$$\Delta CDF = CDF_{app} - CDF_{base}$$

  Equation 7-5

$$\Delta CDF_{j}^{+} = CDF_{j,app}^{+} - CDF_{j,base}^{+}$$

  Equation 7-6

Similar to the approach for Case 1a in Section 7.3. 11, ΔCDF^{+}_{j} will be used in the calculation of RAW values to determine the importance of a change to a given basic event. Equations 7-5 and 7-6 provide a means of assessing the potential vertical movement of a risk metric difference into unacceptable regions of the RG 1.174 acceptance guideline diagram. Any combination of CDF and ΔCDF (or LERF and ΔLERF) that resides in Region I of the diagrams in Figure 7-4 would qualify the source of model uncertainty or related assumption as being potentially key to the application.

In performing the sensitivity analysis, the sources of model uncertainty and related assumptions are linked to the four main ways in which model uncertainty can impact a PRA:

- a single basic event (Case 2a)
- multiple basic events (Case 2b)
- the logic structure of the PRA (Case 2c)
- logical combinations (Case 2d)

7.3.2.1 Case 2a: Sources of Model Uncertainty and Related Assumptions Linked to a Single Basic Event

The sources of model uncertainty and related assumptions identified in Step E-1 are reviewed to determine those that are relevant only to a single basic event. For each identified source of uncertainty, a conservative screening or a realistic sensitivity analysis is performed.

Conservative Screening Option

In Equation 7-5, CDF_{base} and CDF_{app} are calculated using the base PRA and the modified application PRA, respectively. In Equation 7-6, the base and modified application PRAs are recalculated with the value of the relevant basic event (or the jth basic event) set to 1 in both the base and modified application PRAs. By quantifying the base and modified application PRAs, the right-hand terms of Equations 7-5 and 7-6 can be calculated to solve for ΔCDF and ΔCDF^+_j, respectively. These metrics, together with their respective base CDFs, are combined to form two ordered pairs as shown in Table 7-2. These ordered pairs are plotted on the RG 1.174 acceptance guidelines diagrams in Figure 7-4 to determine the acceptability of the application.

Table 7-2 Ordered pairs of CDF and ΔCDF and comparison against acceptance guidelines.

Ordered pair	Purpose
$(CDF_{base}, \Delta CDF)$	Comparison of the mean CDF and mean ΔCDF against the acceptance guidelines. Provides the analyst=s best judgment of the impact of the change in risk.
$(CDF^+_{j,base}, \Delta CDF^+_j)$	Comparison of the greatest possible shift in the base CDF and the greatest possible shift in the ΔCDF, as defined with the jth basic event quantified as 1, against the acceptance guidelines. Provides a perspective on the potential shift in both the ΔCDF and CDF value resulting from an alternate model or assumption.

A source of model uncertainty or related assumption can challenge an application's acceptance guidelines by moving the ordered pair into a different region of Figure 7-4 or close to the boundary of a new region. Sources of model uncertainty and related assumptions associated with the base PRA can impact both CDF_{base} and ΔCDF, which affects the horizontal and vertical position of the ordered pair. If a source of model uncertainty is unique to the change purposed in the application only ΔCDF could change since the model uncertainty would not impact the base PRA. Thus, only the vertical position of the ordered pair is affected.

If the ordered pair, associated with the source of model uncertainty, were to lie within a region of the acceptance guideline (or close to such a region) that could affect the regulator's decision, the issue is potentially key and should be assessed with a sensitivity analysis. Examples of an impact on the application include rejecting the application if the result moves into Region I or introducing compensatory measures if the sensitivity study moves the result from Region III into Region II.

The significance of the ordered pair ($CDF^+_{j, base}$, ΔCDF^+_j) is that it gives a perspective of the potential impact of a model uncertainty issue on both the base PRA and the modified application PRA, whereas the ordered pair (CDF_{base}, ΔCDF) illustrates the impact of parameter uncertainty on both the base and application PRAs, but not the potential impact of model uncertainties.

Another method to assess the potential impact of a model uncertainty relevant to a single basic event is the method of Reinert and Apostolakis [Reinert, 2006]. Reinert and Apostolakis employ a method in which they define the concept of a "threshold RAW" value (analogous to the use of RAW_{max} in Case 1a) for basic events with regard to both CDF and ΔCDF. Their definition of RAW, with regard to CDF, is taken directly from the state of practice of PRA:

$$RAW_{j,CDF-base} = \frac{CDF^+_{j,base}}{CDF_{base}}$$

Equation 7-7

$$RAW_{j,CDF-app} = \frac{CDF^+_{j,app}}{CDF_{app}}$$

Equation 7-8

RAW with regard to ΔCDF is defined as:

$$RAW_{j,\Delta CDF} = \frac{\Delta CDF^+_j}{\Delta CDF}$$

Equation 7-9

where ΔCDF and ΔCDF^+_j are defined as in Equations 7-5 and 7-6, respectively. Substituting Equations 7-5 and 7-6 into Equation 7-9 yields:

$$RAW_{j,\Delta CDF} = \frac{CDF^+_{j,app} - CDF^+_{j,base}}{CDF_{app} - CDF_{base}}$$

Equation 7-10

Solving the relationships in Equations 7-7 and 7-8 for $CDF^+_{j,base}$ and $CDF^+_{j,app}$, respectively, and inserting the results into Equation 7-10 yields:

$$RAW_{j,\Delta CDF} = \frac{[RAW_{j,CDF-app} \times CDF_{app}] - [RAW_{j,CDF-base} \times CDF_{base}]}{CDF_{app} - CDF_{base}}$$

Equation 7-11

The right-hand terms of Equation 7-11 are readily calculable by quantifying both the base PRA and the modified PRA, which allows the analyst to calculate a $RAW_{j,\Delta CDF}$ value directly from results of the base and modified PRA. Reinert and Apostolakis use the relationships in Equations 7-9 and 7-11 to calculate threshold RAWs with regard to both CDF and ΔCDF by selecting maximum acceptable values for CDF and ΔCDF (which they refer to as $CDF_{threshold}$ and $\Delta CDF_{threshold}$, respectively) and then substitute these threshold values for $CDF^+_{j,base}$ in Equation 7-7 and ΔCDF^+_j in Equation 7-9 to yield:

$$RAW_{CDF,threshold} = \frac{CDF_{threshold}}{CDF_{base}}$$

Equation 7-12

NUREG-1855, Revision 1

$$RAW_{\Delta CDF,threshold} = \frac{\Delta CDF_{threshold}}{\Delta CDF}$$ Equation 7-13

where $CDF_{threshold}$ is the value of CDF that corresponds to the vertical line between the applicable regions in Figure 7-4, and $\Delta CDF_{threshold}$ is the value of ΔCDF that corresponds to the horizontal line between the applicable regions in Figure 7-4.

Equation 7-12 and 7-13 yield threshold values for the RAW with regard to the base PRA CDF defined in Equation 7-7 and the RAW with regard to the modified PRA CDF defined in Equation 7-8. The base PRA and the modified PRA are quantified to calculate $RAW_{j, CDF-base}$ values for all basic events in the relevant parts of the base PRA and to calculate $RAW_{j, CDF-app}$ for all basic events in the modified PRA. This allows for the solving of Equation 7-11 to calculate $RAW_{j, \Delta CDF}$ values. The resulting values for $RAW_{j,CDF-base}$ and $RAW_{j,\Delta CDF}$ are compared to the threshold values calculated by Equation 7-12 and 7-13 to determine if any model uncertainty associated with a single basic event poses a potential key model uncertainty.

In using the method of Reinert and Apostolakis, care should be given to assessing the potential combined impact of a model uncertainty on both CDF and ΔCDF. This method does not automatically investigate the potential that, when taken separately, the CDF and ΔCDF could satisfy the acceptance guideline for the application but, when taken together, could result in an overall unacceptable result. This is the function of the order pairs $(CDF_{base}, \Delta CDF)$ and $(CDF^+_{j,base}, \Delta CDF^+_j)$ in the ordered pair approach discussed above. Reinert and Apostolakis do address this issue by selecting more than one threshold value for CDF and ΔCDF based on the horizontal and vertical transitions between Regions I and II and between Regions I and III.

Reinert and Apostolakis provide a case study to illustrate this method.

Realistic Sensitivity Assessment Option

The terms for the ordered pairs in Table 7-2 are evaluated for any credible hypothesis developed for any source of model uncertainty or related assumption linked to the jth basic event. For any such credible hypothesis, if any of the ordered pairs in Table 7-2 yields a result in or close to Region I, then the source of model uncertainty or related assumption is a key uncertainty.

As discussed above, Reinert and Apostolakis provide an alternate method to test whether a source of model uncertainty or related assumption linked to a single basic event constitutes a potential key uncertainty. To perform a sensitivity analysis on model uncertainty issues that can be related to specific basic events, Reinert and Apostolakis continue to employ the concept of a threshold RAW. The term "threshold" RAW importance measure coined by Reinert and Apostolakis also can be thought of as a maximum acceptable RAW importance measure. The maximum acceptable RAW represents the largest possible value for the RAW importance measure for which it would be mathematically impossible for the uncertainty associated with a particular basic event to produce a PRA result that would move from one region of the figure to another. Threshold values for the RAW_{cdf} and the $RAW_{\Delta cdf}$ are calculated as shown in Equation 7-12 and 7-13.[32]

---

[32] These are Equations 12 and 13 in Reinert, 2006.

Once the $RAW_{CDF, threshold}$ and the $RAW_{\Delta CDF, threshold}$ have been calculated, each model issue can be evaluated by investigating the model for the basic events relevant to each issue. For any particular issue, the value for a particular relevant basic event j is adjusted, and both the base PRA and the modified PRA are reevaluated until a result for the base PRA is found that yields one of the following:

$$RAW_{j,CDF} \approx RAW_{CDF,threshold}$$ Equation 7-14

$$RAW_{j,\Delta CDF} \approx RAW_{\Delta CDF,threshold}$$ Equation 7-15

If the value of the jth basic event that corresponds to the approximation in either Equation 7-14 or 7-15 is based on a credible hypothesis for the basic event's probability, then the source of model uncertainty or related assumption linked to the jth basic event is a key uncertainty. If this is the case, the sensitivity analysis should continue so that a maximum probability value for the jth basic event and its corresponding high estimates for CDF and ΔCDF are calculated. These high estimates for CDF and ΔCDF will most likely be less than the values of $CDF^+_{j, base}$ and ΔCDF^+_j, respectively, that were calculated by setting the value of the jth basic event to 1. These high values for CDF_j and ΔCDF_j will be necessary for the comparison of the risk-informed application to the RG 1.174 acceptance guideline. If the value of the jth basic event that yields the approximation in either Equation 7-14 or 7-15 is based on a hypothesis that is not credible, then the issue is not a key uncertainty. The uncertainty associated with the issue could not affect the application's results to the point that the results either exceed or challenge the application's acceptance guidelines.

7.3.2.2 Case 2b: Sources of Model Uncertainty and Related Assumptions Linked to Multiple Basic Events

The sources of model uncertainty and related assumptions identified in Step E-1 are reviewed to determine those that are relevant to multiple basic events. An example would be the choice of model to quantify human errors and recovery actions, or an assumption that affects the quantification of a particular failure mode of several redundant components (e.g., stuck open safety relief valve). For each identified source of uncertainty, a conservative screening or a realistic sensitivity analysis is performed.

Conservative Screening Option

The RAW importance measures for a given group of basic events, which share a common source of model uncertainty or related assumption, cannot be used to collectively assess the combined impact on the application of that source of model uncertainty or related assumption (i.e., adding together the individual RAW importance measures of the basic events in a group does not result in a "group" RAW importance measure). As such, to determine the true impact on the risk model of the model uncertainty or related assumption associated with that group of events, the basic event probabilities for that group must all be set to a value of 1, and the entire PRA model must be requantified. The ordered pairs in Table 7-2 can now be calculated, where similar to the case in Section 7.3.1.2, the subscript j becomes k to represent the set of basic events relevant to the kth source of model uncertainty. If the ordered pair associated with the source of model uncertainty were to lie in a region of the acceptance guideline (or close to such a region) that could affect the regulator's decision, the issue is potentially key and should be assessed with a sensitivity analysis.

Realistic Sensitivity Assessment Option

The analyst selects credible alternative models or assumptions for each source of model uncertainty or related assumption associated with multiple basic events. For each credible alternative model, the base and modified application PRAs are requantified using basic event probability values generated from the alternative models. The result is a set of new CDF and ΔCDF ordered pairs, which represent the range of potential PRA results, each of which needs to be compared against the RG 1.174 acceptance guidelines. If any one of the results produces an ordered pair that challenges or exceeds the acceptance guideline in Figure 7-4, the issue is considered key to the application.

7.3.2.3 Case 2c: Sources of Model Uncertainty and Related Assumptions Linked to the Logic Structure of the PRA

The sources of model uncertainty and related assumptions identified in Step E-1 are reviewed to determine those that are relevant to the logic structure of the PRA. For each identified source of uncertainty, a conservative screening or a realistic sensitivity analysis is performed.

Alternative methods or assumptions need to be assessed by manipulating or altering the PRA model to reflect those alternatives. This includes alternative methods or assumptions that could possibly introduce the following:

- new cut sets in existing sequences by changing fault tree models
- new sequences by changing the structure of event trees
- new classes of accident sequences by introducing new initiating events

New estimates for $CDF^+_{n,base}$ and $CDF^+_{n,after}$ can be developed, where these terms are defined as follows:

$CDF^+_{n,base}$ The base PRA CDF mean estimate where the base PRA has been modified to address the nth source of model uncertainty or related assumption that is linked to the logic structure of the PRA.

$CDF^+_{n,app}$ The base PRA CDF mean estimate where the PRA, as modified for the application, has been further modified to address the nth source of model uncertainty or related assumption that is linked to the logic structure of the PRA.

Conservative Screening Option

The effort to change the PRA logic structure can involve significant resources. However, in some cases, it may be possible to perform an approximate bounding evaluation (see Section 5.2, Step C-1) that can demonstrate that the potential impact of the alternate assumption or model will not produce a result that challenges the application's acceptance guidelines. As an example, this demonstration can be achieved if the effect of the model uncertainty or related assumption is limited to the later branches of the lower frequency accident sequences and the frequency of the portion of the sequences up until the branch points is low enough.

Realistic Sensitivity Assessment Option

The analyst selects credible alternative models or assumptions for the particular issue. Then, for each credible alternative model, the base and modified PRAs are requantified. Using Equation 7-5 and 7-6, the analyst can calculate values for the terms of the ordered pairs in Table 7-2 and compare the plots of those ordered pairs to the acceptance guidelines shown in Figure 7-4. If the ordered pair associated with a source of model uncertainty were to lie in a region (or close to such a region) of the acceptance guideline, the issue is potentially key to the application.

7.3.2.4 Case 2d: Sources of Model Uncertainty and Related Assumptions Linked to Logical Combinations

The sources of model uncertainty and related assumptions identified in Step E-1 are reviewed to determine those that are relevant to combinations of basic events and logic structure. One should not, however, restrict oneself to a short list of generic logical groupings. The analyst's judgment and insight regarding the PRA should yield logical groupings specific to the PRA in question. Certain issues may readily fall into more than one logical grouping depending on the nature of the other issues. For these cases, the combination may impose a synergistic impact on the uncertainty of the PRA results. Section 4.3.2 of EPRI report 1016737 [EPRI, 2008] and Section 4.4.2 of EPRI report 1026511 [EPRI, 2012] present further guidance on grouping issues into logical groups. See the discussion for Case 1d in Section 7.3.1.4 for examples.

Conservative Screening Option

When all the contributors to the logical group of sources of model uncertainty impact only basic events, the approach is similar to Case 2b with regard to quantitative screening. The concept of setting all relevant basic events to 1 simultaneously and then reevaluating the PRA yields the same perspective for a logical grouping of sources of uncertainties as for a single source of uncertainty that impacts several basic events. Hence, all basic events relevant to a particular model issue or to a logical group of issues are set to 1 to calculate the ordered pairs in Table 7-2, where the index becomes k to represent the set of basic events relevant to the kth logical group of model issues. If none of the sensitivity cases associated with the logical group could impact the application's results by moving the results from one region of the figure to another (or close to the boundary), for example, then the sources of uncertainly included in the logical grouping do not present potential key sources of model uncertainty. No mathematical possibility exists that the values of the basic events linked to the particular logical grouping could achieve an unacceptably high CDF. However, should at least one sensitivity case move the PRA's result into or close to another region, the sources included in the logical group of sources of model uncertainty are potential key and should be evaluated with a realistic sensitivity analysis.

If the logical combinations involve impacts on both the basic event values and the PRA structure, the process of performing conservative screening becomes more involved—as for Case 1d (Section 7.3.1.4). The impacts on the PRA structure should be assessed first so that the impacts on the basic event probability values can then be assessed with the modified logic structure. The approach described in Case 2c (Section 7.3.3.3) can be used to address the effects of the sources of model uncertainty on the PRA logic. The modified PRA structure can then be used in conjunction with the process in Case 2b (Section 7.3.3.2) to assess the effect of a credible alternative model on multiple basic events, as evaluated using the modified PRA structure.

Realistic Sensitivity Assessment Option

The analyst selects credible alternative models or assumptions for each logical group of sources and assumptions and the PRA logic and basic event probability values should be modified based on the selected alternative model; the logic is modified first and then the basic event probabilities. For each credible alternative model, the base and modified application PRAs are requantified using basic event probability values generated from the alternative models. The result is a set of new CDF and ΔCDF ordered pairs, which represents the range of potential PRA results. Each pair needs to be compared against the RG 1.174 acceptance guidelines. If any one of the results produces an ordered pair that exceeds or challenges the application's acceptance guidelines in Figure 7-4, the issue is considered key to the application.

7.4   Summary of Stage E

Stage E has two parallel and similar (but not exactly identical) paths as shown in Figure 7-1. The path through Steps E-1 and E-2 represents the process that is followed when the risk results calculated in Stage D for an application meet the application acceptance guidelines. The path through Steps E-3 and E-4 represents the process that is followed when the risk results calculated in Stage D do not meet the application's acceptance guidelines.

The goal is to ultimately determine whether (and the degree to which) the quantitative acceptance guidelines are impacted by sources of model uncertainty and related assumptions. The process of addressing model uncertainty corresponds to Stage E of the overall process for the treatment of uncertainties. At the completion of Stage E, the process continues to Stage F (Section 8) regardless of whether the estimate of the risk metrics calculated in Stage D (Section 6) challenge or exceed the acceptance guidelines, In Stage F the applicant develops a strategy to address any key uncertainties and propose any compensatory measures or performance monitoring programs as appropriate, and then prepares documentation for presenting the application to the NRC for a decision, which is discussed in Stage G (Section 9).

8.   STAGE F — LICENSEE APPLICATION DEVELOPMENT PROCESS

This section provides guidance to the licensee on the process of developing a risk-informed application submittal, as the process relates to the treatment of uncertainties associated with the application probabilistic risk assessment (PRA). The purpose of Stage F is to help ensure that adequate justification is provided for the acceptability of the risk-informed application. Further, the guidance for this stage helps ensure that the argument for adequate justification is included in the documentation clearly and concisely.

8.1   Overview of Stage F

Stage F consists of a set of options that are used by the licensee throughout the uncertainty assessment process to ensure that the application risk results meet the acceptance guidelines or that adequate justification is provided for the acceptability of the results. When in this Stage, the licensee may perform one or more of the following three options: (1) redefine the application; (2) refine the PRA analysis; or (3) use compensatory measures or performance monitoring requirements. For example, if the results of a parameter or model uncertainty assessment challenge or exceed the acceptance guidelines and that challenge or exceedance cannot initially be justified, the licensee can use the options in this stage to either reduce the amount of justification needed (i.e., redefining the application or refining the PRA) or to provide adequate justification for the acceptability of the challenge or exceedance (i.e., compensatory measures, performance monitoring). The relationship of Stage F to the overall uncertainty assessment process is illustrated below in Figure 8-1.

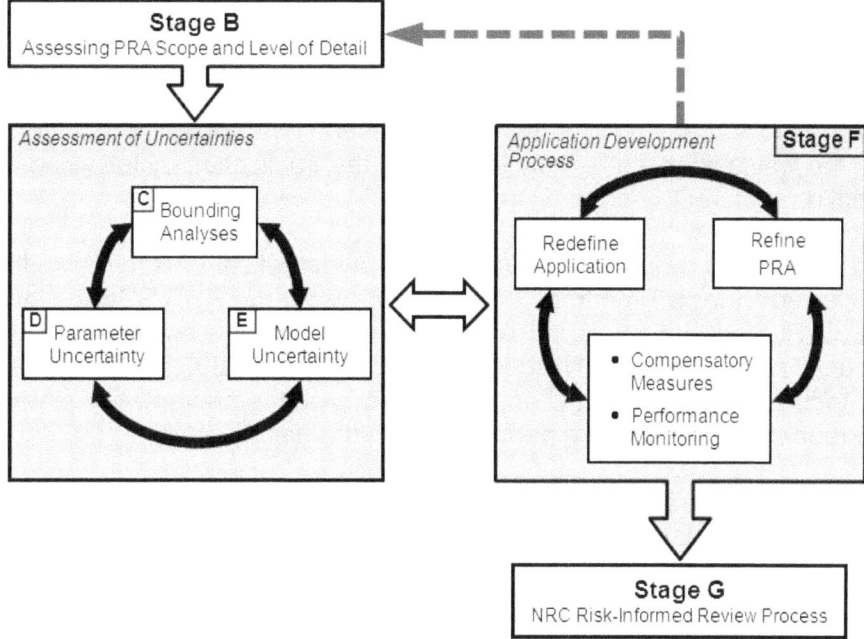

Figure 8-1 Overview of Stage F and its relationship to the process of assessing uncertainties

The left gray box in Figure 8-1 represents the process of assessing uncertainties and uses the output of Stage B as an input to Stages C, D, and E. The right gray box represents Stage F and the three options that may be used to develop an application. The output of Stage F is a risk-informed application that is ready to be submitted to the NRC. The three options in Stage F may be used separately or together, and to varying degrees, when assessing the impact of uncertainties for a particular application, consequently Stage F is performed in an iterative fashion. Both of the first two options direct the licensee back to previous stages of the uncertainty assessment process to reassess the impact of uncertainties after either redefining the application or refining the PRA analysis. Specifically, when refining the PRA analysis, the licensee must return to Stage D and then Stage E to recalculate the risk metrics, reassess the impact of model and parameter uncertainties, and determine the acceptability of the new risk measure values relative to the acceptance guidelines. When redefining the application, the licensee must return to Stage B—as indicated by the dashed gray line—to determine scope of the redefined application and subsequently reassess the impact of all uncertainties via Stages C, D, and E. The third option, using compensatory measures and performance monitoring requirements, directs the licensee to produce additional justification to achieve the needed level of justification for a given application. Before explaining more about the three options, a general discussion is provided on the relationship of Appendix A to the guidance provided in this report.

8.2   Example of the Licensee Application Development Process

An illustrative example of the whole licensee development process is presented in EPRI report 1026511 [ERPI, 2012]. The license amendment request (LAR) discussed in EPRI report 1026511 is a hypothetical example developed to illustrate the process described both this NUREG and in EPRI reports 1026511 and 1016737 [EPRI, 2008]. Although the example is hypothetical, a realistic PRA model has been used. The PRA model has been modified to ensure that there are some sources of uncertainty that illustrate several aspects of the process discussed in the previous sections of this report. As pointed out earlier, the stages and steps discussed in this report may be iterative and may vary in the order of their implementation. While the example is meant to illustrate the implementation of the guidance given in this report, it does not follow exactly the order of the steps as presented in the report. Rather, the example is presented in the way a licensee is likely to develop the application by following the process discussed in this report.

The example risk-informed regulatory application is a hypothetical LAR to revise the Technical Specification Allowed Outage Time from 3 days to 7 days for the RHR/SPC system at a representative BWR, Mark II plant. The PRA model for the plant is consistent with the PRA technical adequacy requirements outlined in Regulatory Guide 1.200 [NRC, 2007a]. The purpose of the technical specification change is to allow routine preventive maintenance currently performed at shutdown to be performed with the unit at power.

No single example can illustrate all parts of the process described in this report, but the Appendix A example does demonstrate many of the aspects of the previously discussed stages. Specifically it illustrates the following facets of the indicated stages:

- Stage A — The LAR is an activity that fits within the guidance provided in this NUREG.

- Stage B — The scope and level of detail of the PRA needed for this application is established. The needed risk metrics and the corresponding acceptance guidelines are identified.

- Stage C — Hazards not modeled in the PRA are either screened from consideration or shown to be not risk significant based on a conservative analysis.

- Stage D — Parameter uncertainties are assessed. The EPRI guidelines in EPRI report 1016737 are implemented to show that for this case the state-of-knowledge correlation (SOKC) is not important for calculating the risk metrics. An initial comparison of the risk metrics with acceptance guidelines shows the acceptance guidelines could be exceeded. The fire PRA part of the analysis is refined to eliminate some demonstrated conservatisms, and some compensatory measures are imposed. The refined results show that the risk measures now meet the acceptance guidelines.

- Stage E — Model uncertainties are assessed and key model uncertainties identified. Sensitivity cases for the key model uncertainties are explored, and where the sensitivity cases challenge the acceptance guidelines, additional potential compensatory measures are identified.

The approach discussed in the EPRI report 1026511 for addressing the process of dealing with uncertainty is provided as one example of a spectrum of possible approaches and is provided for discussion and illustrative purposes. As is stated throughout the example, it should not be construed to imply that this is the only approach or that the specifics of the illustrative example would be sufficient in all cases. It should also be noted that it is not the intent of this example to imply that compensatory measures would be required for every application, but rather that when there are key sources of uncertainty and/or the acceptance guidelines are being challenged, then compensatory measures are one acceptable means of providing additional confidence to the decision maker.

8.3 Redefine the Application

This section provides guidance on redefining the risk-informed application. When a PRA model is incomplete in its coverage of significant risk contributors, the scope of a risk-informed application can be restricted to those areas supported by the risk assessment. For example, if the PRA model does not address fire hazards, the change to the plant could be limited such that any structures, systems, or components (SSCs), which are used to mitigate the risk from fires, would be unaffected. In this way, the contribution to the overall risk from internal fires would be unchanged. This is the strategy adopted in Nuclear Energy Institute (NEI) report 00-04 [NEI, 2005b] for categorizing SSCs according to their risk significance when the PRA being used does not address certain hazard groups.

If the risk-informed application is redefined, the licensee should subsequently perform Stage B to reassess the scope and level of detail needed for the redefined application. The licensee should then perform Stages C, D, and E again to determine if and how the impact of the completeness, parameter, and model uncertainties has changed for the redefined application.

8.4 Refine the PRA Analysis and Recalculate Risk Measures

This section provides guidance on refining the application PRA analysis and recalculating the risk metrics. If the PRA analysis is refined (e.g., the level of detail is increased to produce more realistic results, model uncertainties that have been introduced by assumptions are removed via the development of more sophisticated models), the licensee should perform Stage D wherein the risk metrics are recalculated with the refined PRA and the impact of parameter uncertainty is reassessed. Next, the licensee should perform Stage E again to determine the impact of model uncertainties and related assumptions on the new risk metrics.

8.5 Use of Compensatory Measures or Performance Monitoring

This section provides guidance on the use of compensatory measures or performance monitoring for the purpose of providing additional justification for the acceptability of the application. Compensatory measures can be used to neutralize the expected negative impact of some feature of plant design or operation on risk. For example, a fire watch may be established to compensate for a weakness identified in the fire PRA such as a faulty fire barrier or the temporary removal of a fire barrier. Another example is the implementation of a manual action to replace an automatic actuation of a system, such as initiating depressurization of a reactor pressure vessel for a boiling water reactor. In the latter example, the compensatory action can be modeled explicitly in the PRA and the impact of the manual action on the risk results can be directly observed by requantifying the PRA and determining the change in risk. However, if the compensatory action is not explicitly modeled in the PRA—as in the case of the fire watch—it is necessary to understand the specific scenarios for which the compensatory measures have been implemented so that the impact of the compensatory measures is well understood. For example, establishing that the types of fires that a fire watch is intended to mitigate are slow growing would add confidence to the value of the fire-watch. On the other hand, if a high-energy arcing fault were a significant contributor to the fires in an area, a fire watch would be ineffective.

Performance monitoring can be used to demonstrate that, following a change to the design of the plant or operational practices, there has been no degradation in specified aspects of plant performance that are expected to be affected by the change. This monitoring is an effective strategy when no predictive model has been developed for plant performance in response to a change. One example of such an instance is the impact of the relaxation of special treatment requirements (in accordance with 10 CFR 50.69) on equipment unreliability. No consensus approach to model this cause-effect relationship has been developed. Therefore, the approach adopted in NEI 00-04 as endorsed in Regulatory Guide 1.201, "Guidelines for Categorizing Structures, Systems, and Components in Nuclear Power Plants According to Their Safety Significance," [NRC, 2006a] is to:

- Assume a multiplicative factor on the SSC unreliability that represents the effect of the relaxation of special treatment requirements.

- Demonstrate that this degradation in unreliability would have a small impact on risk.

Following acceptance of an application which calls for implementation of a performance monitoring program, such a program would have to be established to demonstrate that the assumed factor of degradation is not exceeded. For monitoring to be effective, the plant performance needs to be measurable in a quantitative way, and the guidelines used to assess acceptability of performance need to be realistically achievable given the quantity of data that might be generated.

Appendix A provides example illustrations of how compensatory measures are demonstrated - through PRA calculations - to strengthen the case for an application, Section A.5 for the PRA modified for the application, and A.7.2 for key uncertainties.

After determining that adequate justification has been developed for the acceptability of the application, the licensee may prepare the application submittal as discussed in the following section.

8.6   Preparing the Application Submittal

This guidance discusses the documentation elements needed for a risk-informed application submittal. The licensee is responsible for ensuring that adequate justification is provided for the acceptability of the application. In doing so, the licensee must ensure that the conclusions of the risk assessment are communicated clearly and concisely and are adequately documented.

An important part of the documentation of the risk assessment is a discussion of the robustness of, or the confidence in, the conclusions drawn from that analysis. When justifying the acceptability of a risk-informed application, the licensee should carefully analyze the technical bases of the application to ensure that the application demonstrates the following:

- A clear understanding of the risk contributors and their impact on the results.

- Model uncertainties have been accounted for using the techniques discussed in Stage E (Section 7)

- The model has sufficient scope and level of detail to support the conclusions of the analysis using the guidance from Stage C (Section 5).

An issue that can result in the need for more detailed documentation is aggregation of risk results (i.e., added together) from different hazard groups. For all applications, it is necessary to aggregate the PRA results from the applicable hazard groups (e.g., from internal events, internal fires, and seismic events). Because the hazard groups and plant operation states are independent, addition of each hazard groups' or plant operating states' (POSs') risk metric results (e.g., core damage frequency, large early release frequency) is mathematically valid. However, it is important to note that, when combining the results of PRA models for several hazard groups - as may be required by certain acceptance guidelines - the PRA level of detail and approximation may differ between hazard groups with some being more conservative than others. Significantly higher levels of conservative bias can exist in PRAs for external hazards, low-power and shutdown operational modes, and internal fire PRAs than for at-power PRAs. These biases result from several factors, including the unique methods or processes and the inputs used in these PRAs as well as the scope of the modeling. For this reason, when aggregating PRA results, it is important to understand both the level of detail associated with the modeling of each of the hazard groups and POSs as well as the hazard group-specific model uncertainties.

For each hazard group individually – consistent with the guidance of Section 7 - the sources of model uncertainty and related assumptions are identified and their impact on the results analyzed so that, when it is necessary to combine the PRA results, the overall results can be characterized appropriately and the results aggregated meaningfully. The understanding gained from these analyses is used to appropriately characterize the relative significance of the contributions from each hazard group. The effects of aggregating risk results become more important as the overall risk metric approaches closer to the acceptance guidelines. For example, the importance of a conservative bias would be much greater when the risk metric approaches, challenges, or exceeds the acceptance guidelines as opposed to when the risk metric is far below the acceptance guideline.

The differing level of detail in the individual hazard and plant operating state analyses is also important when considering risk ranking or categorization of SSCs. For applications that use risk importance measures to categorize or rank SSCs according to their risk significance (e.g., revision of special treatment), a conservative treatment of one or more of the hazard groups can bias the final risk ranking. Moreover, the importance measures derived independently from the analyses for different hazard groups cannot be simply added together and thus would require a true integration of the different risk models to evaluate them.

To facilitate the review of an application with aggregated results, the results and insights from all of the different risk contributors relevant to the application should be provided explicitly in the application in addition to the aggregated results.

8.6.1 Documentation Elements

The purpose of this section is to provide guidance on the documentation needed for an application submittal. In general, the quality of the justification needed to demonstrate the acceptability of an application increases depending on whether the application risk measures estimates approach, challenge, or exceed the acceptance guidelines. The quality of the documentation that needs to be developed for the application will depend on the justification needed for the acceptability of the application and will vary from one application to another. In general, the documentation in the application submittal should consist of the following elements:

- A description of the PRA scope used in the application and applicability of the scope to the application. This should include the specific risk calculations that are derived from evaluating the PRA model for the application. This description should also include an explanation of the rationale behind excluding of any part of the base PRA scope.

- A description of the acceptance guidelines that are used for comparison with the risk metrics.

- A discussion of the adequacy of the treatment of uncertainties. For example, this may include an explanation of how the treatment of uncertainties meets the ASME and American Nuclear Society PRA standard [ASME/ANS, 2009], a summary of positive findings from the peer review of the PRA that relate to uncertainties, and an explanation of how the application follows the process in NUREG-1855. If NUREG-1855 was not used, the application should include a description of the process used and an explanation of how it is equivalent to NUREG-1855.

- A discussion of the impact of the parameter uncertainty on the risk metrics.

- A description of the relevant sources of model uncertainty and their impact on the results.

- A description of any significant modeling assumptions.

- A statement justifying that the application risk results are acceptable and that the application should be accepted with or without compensatory measures. When the application is characterized as Regime 1, it is generally sufficient to provide a qualitative argument describing why parameter uncertainty and model uncertainty do not impact the decision.

Although the documentation needed for an application is described in terms of general elements, these elements should be developed to the degree necessary to provide a clear and concise presentation of the argument that adequate justification has been provided for the acceptability of the risk-informed application.

8.7   Summary of Stage F

This section provides guidance to the licensee on the process of developing a risk-informed application submittal, as related to the treatment of PRA uncertainties. Stage F describes the process used by the licensee to ensure that (1) adequate justification has been provided for the acceptability of the risk-informed application and (2) that the documentation provides a clear and concise presentation of this argument.

After the risk-informed application has been developed, it is submitted to the NRC for the staff's risk-informed review. This review represents Stage G, the last stage in the process of assessing the treatment of uncertainty associated with a PRA in a risk-informed application.

Appendix A of this report provides an example of the implementation of the guidance in this report and follows the process discussed in this section.   As discussed, the process of developing an application is iterative in nature and will not necessarily follow the guidance in this report in the order in which it is presented. Similarly, the example illustrates the implementation of the guidance given in this report and demonstrates the iterative nature of the application development process.

9.   STAGE G — NRC RISK-INFORMED REVIEW PROCESS

The purpose of this section is to describe the process used by the staff to determine whether a licensee's risk-informed application demonstrates an acceptable treatment of uncertainties and that the proposed application represents an acceptable risk impact to the plant. The staff's risk-informed review is the last stage in the process of assessing the treatment of uncertainties associated with a probabilistic risk assessment (PRA) in a risk-informed application.

9.1   Staff Overall Review Approach

The risk-informed review process is comprised of several steps, not all of which relate to the treatment of uncertainties. This guidance describes only those aspects of the staff's risk-informed review process that specifically relate to the treatment of uncertainties. Guidance on the other aspects of the staff's risk-informed review process may be found in other application-specific guidance documents and is briefly discussed later in this section.

The staff review of a risk-informed application begins with the comparison of the application risk results to the acceptance guidelines. The justification needed to demonstrate the acceptability of a given risk-informed application is largely dictated by the proximity of the risk results to the acceptance guidelines. In general, an application can be characterized as falling into one of the following four general regimes based on the proximity of the risk results to the acceptance guidelines:

- Regime 1—The risk results are well below the acceptance guidelines
- Regime 2—The risk results are closer to, but do not challenge the acceptance guidelines
- Regime 3—The risk results challenge the acceptance guidelines
- Regime 4—The risk results clearly exceed the acceptance guidelines

Figure 9-1 illustrates how these four regimes relate to the comparison of the application risk results and the acceptance guidelines.

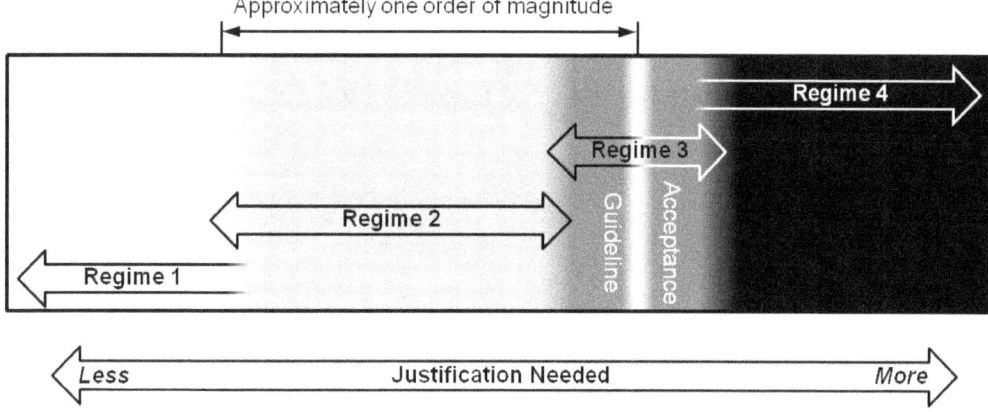

Figure 9-1 Relationship of the comparison regimes to the acceptance guidelines

The justification for a given application should be commensurate with the proximity of the risk results to the acceptance guidelines, as shown above. In general, more justification will be needed for a given application when the risk results are closer to challenging or exceeding the acceptance guidelines than when the risk results are further below the acceptance guidelines.

In determining whether the acceptance guidelines have been met, the staff seeks to answer the following general questions:

- How do the risk results compare to the acceptance guidelines?
- Is the scope and level of detail of the PRA appropriate for the application?
- Is the PRA model technically adequate?
- Is the acceptability of the application adequately justified?

Similar to the licensee's development of the risk-informed application, the staff's risk-informed review process is not necessarily performed in a serial manner, nor is the transition from one portion of the review process to another always absolute. The staff's risk-informed review is a dynamic process that often relies on additional information beyond the original application submittal that the NRC may request from the licensee. In general, when the staff makes a determination in a given step of the process, the determination is based on a review of the submittal documentation in conjunction with any information received via open and continuous dialogue with the licensee. This dialogue is meant to achieve the clearest understanding of the application and generally consists of oral discussions and written correspondence. This dialogue may also result in the generation of official requests for additional information by the staff, all of which are formally documented and considered together with the original submittal. In this way, the staff considers the original submittal documentation and any additional information and insights gained from the review process as a whole during the risk-informed review of the licensee's application.

9.2   Risk-Informed Review of Completeness Uncertainty

The purpose of this section is to describe the process the staff uses to determine whether a licensee's treatment of completeness uncertainty associated with the PRA is acceptable for a given risk-informed application. In making this determination, the staff assesses whether the following criteria are met:

- The PRA scope and level of detail and the licensee's use of any screening analyses are appropriate for the application.

- The base and revised PRA that are used to support the application are technically adequate.

Figure 9-2 illustrates the overview of the completeness uncertainty review process.

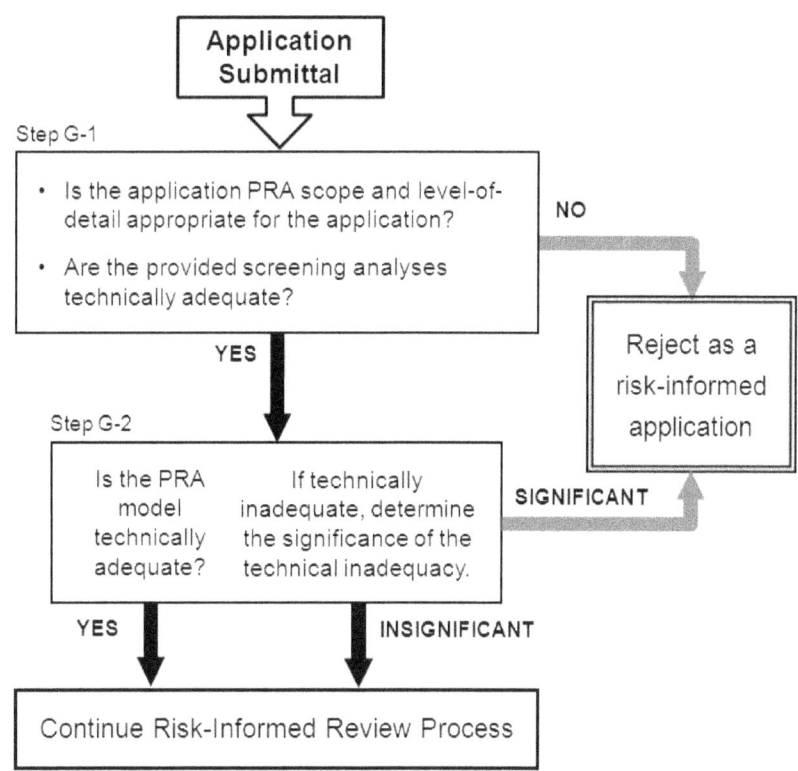

Figure 9-2 Overview of the risk-informed review of completeness uncertainty

As shown above, the staff first determines whether scope and level of detail are appropriate for the application. Next, the staff reviews the base and application PRA to determine if they are technically adequate and then determines whether any identified inadequacies are significant. As described in Section 5, qualitative and quantitative screening analyses are the primary tools used to develop the scope and level of detail of an application. As such, the following discussion focuses primarily on the staff's review of the licensee's screening analyses used in the risk-informed application.

Step G-1: Assessment of Scope, Level of Detail, and the Use of Screening Analyses

The first step performed by the staff in the completeness uncertainty review involves determining whether the needed PRA scope and level of detail in the licensee's application is appropriate for the risk-informed activity being performed. For example, with regard to PRA scope, the staff determines whether the application PRA adequately addresses the hazards and plant operating states needed for the risk-informed activity. With regard to the level of detail in the PRA, the staff may determine, for example, whether the application provides the appropriate level of model detail in the logic models.

If the staff determines that scope and level-of-detail items are missing from the application PRA that are required for the risk-informed activity, the staff reviews the submittal to determine how the licensee addressed those missing elements. If the licensee does not provide adequate justification for the exclusion of the missing PRA scope and level-of-detail items (e.g., the licensee provides no or insufficient screening analyses) the staff will likely reject the licensee's application as a risk-informed application. However, if the licensee provides screening analyses to address the missing PRA scope and level-of-detail items, the staff reviews those screening analyses to determine whether they provide adequate justification. Further, the staff review involves determining whether the screening analyses themselves are technically adequate and whether the licensee performed them in a technically correct manner.

Screening analyses used in PRAs are either qualitative or quantitative. Regarding qualitative screening analyses, the staff review involves determining whether the analysis adequately demonstrates that the screened PRA scope or level-of-detail item of interest cannot affect the decision (i.e., the application) under consideration. Regarding quantitative screening analyses, there are three types of analyses that the staff may encounter: bounding, conservative, and realistic. The staff's determination of whether a quantitative screening analysis provides adequate justification for a missing scope or level-of-detail item is based in part on whether the degree of conservatism employed in the analysis is appropriate for that missing PRA item. In making this determination, the staff considers the importance of the missing PRA item relative to the decision under consideration. In practice, the staff assesses whether the assumptions, models, data, and level of detail used in the screening analysis adequately support screening the PRA item.

- Bounding Analyses: When the licensee has used a bounding analysis to screen a risk contributor, the staff review involves determining whether the analysis adequately demonstrates that the worst credible outcome—of the set of outcomes associated with the risk assessment of the screened item—has been considered. As discussed in Section 5.2.2, the worst credible outcome is the one that has the greatest impact on the defined risk metric(s). As such this determination involves assessing whether the bounding analysis is bounding in terms of the potential outcome and the likelihood of that outcome. Further, the staff review involves determining whether the licensee's analysis adequately demonstrates that the worst credible outcome has been shown to have a negligible impact on the application risk results, in terms of its consequences and likelihood.

- Conservative Analyses: When the licensee has used a conservative, but less-than-bounding analysis to screen a risk contributor, the staff review involves determining whether the analysis adequately demonstrates that all potential impacts, and the effects of those impacts, have been considered. In particular, the staff review involves determining whether any potential impacts or effects of potential impacts, which could lead to a more severe credible outcome, may not have been considered. Further, the staff review involves determining whether the licensee's analysis adequately demonstrates that the frequency of the identified potential impacts is greater than or equal to the maximum credible collective frequency of the spectrum of impacts analyzed for the screened PRA item.

- Realistic Analyses: When the licensee has used a realistic analysis to screen a risk contributor, the staff review involves determining whether the degree of realism incorporated into the analysis is appropriate, relative to the screening criteria used in the analysis. For example, if the screening criteria for a given analysis requires a high degree of realism (i.e., the screening criterion is very detailed and specific), the degree of realism used in the analysis should be appropriately high (e.g., best estimate models and data are used to represent the missing PRA item) so as to produce a more accurate representation of the risk contributor. Thus, for realistic screening analyses, the staff review involves (1) determining the degree of realism dictated by the screening criteria, and (2) determining whether the degree of realism incorporated into the licensee's analysis is commensurate with that which is required.

The staff review of the screening analyses also involves determining whether the screening analyses are technically adequate (i.e., the analyses have been performed in a technically correct manner). This may be accomplished in part by determining how well the analyses address the guidance provided in Section 5 of this report (see Section 5.1.4 for examples of screening analyses).

If the staff determines that the screening analyses do not provide adequate justification for the exclusion of the missing PRA scope and level-of-detail items or that the analyses are technically inadequate, the submittal may be rejected as a risk-informed application. If the screening analyses do provide adequate justification for the exclusion of the missing PRA scope and level-of-detail items and the analyses are found to be technically adequate, the staff review proceeds to Step G-2.

Step G-2: PRA Model Technical Adequacy

The next step in the completeness uncertainty review involves determining whether the application PRA model itself is technically adequate. The guidance used by the staff for this step of the process is contained in application-specific regulatory guides and is not discussed here, as it is beyond the scope of this report. The technical adequacy of the base PRA is addressed in Regulatory Guide 1.200, Revision 1, "An Approach for Determining the Technical Adequacy of Probabilistic Risk Assessment Results for Risk-Informed Activities" [NRC, 2007a]. If the staff determines that the application PRA model is technically adequate, the staff continues the risk-informed review process. If a technical inadequacy is identified, the staff then determines whether the technical inadequacy is significant.

The guidance used by the staff to determine the significance of a technical inadequacy in the application PRA is also contained in application-specific regulatory guides and is not discussed here, as it is beyond the scope of this report. If the staff determines that a technical inadequacy is significant, the application may be rejected as a risk-informed application. If the technical inadequacy is determined not to be significant, the staff continues with the risk-informed review process.

9.3 Risk-Informed Review of Parameter Uncertainty

The purpose of this section is to describe the process the staff uses to determine whether a licensee's treatment of parameter uncertainty associated with the PRA is acceptable for a given risk-informed application. The staff makes this determination based on whether adequate justification has been provided for the acceptability of the risk metrics, as compared to the acceptance guidelines. The staff uses three essential pieces of information from the application submittal in making this determination:

1. an estimate of the relevant risk measure(s), usually expressed as a mean value(s)
2. the probability distribution(s) of the risk measure(s)
3. the acceptance guidelines used for the particular application

Figure 9-3 illustrates the detailed steps of the staff's parameter uncertainty review process.

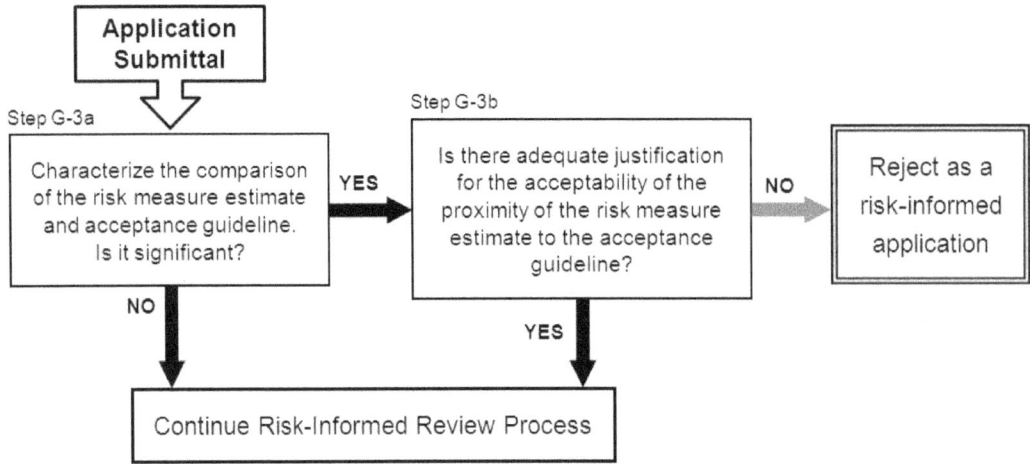

Figure 9-3 Overview of the risk-informed review of parameter uncertainty

As shown above, the staff first characterizes the numerical difference between the risk metric and the acceptance guideline and determines whether the difference is significant. If the difference is not significant, the staff continues on with the risk-informed review process. If the difference is significant, the staff determines whether adequate justification has been provided for the acceptability of the difference. The adequacy of the justification for the difference may not be fully evident until the staff has reviewed the licensee's treatment of model uncertainties. As such, the staff may refrain from rejecting the application as a risk-informed application—on the basis of the treatment of parameter uncertainties alone—until the review of the licensee's treatment of model uncertainties has been performed.

Step G-3a: _Characterize the Comparison of the Risk metric to the Acceptance Guideline_

The first step performed by the staff in the parameter uncertainty review involves characterizing the comparison of the risk metric to the acceptance guideline and determining whether the comparison is significant. As discussed Section 5, a risk metric can be described in terms of a point estimate wherein the parameter uncertainty on this result is represented by a range of values that the actual value may assume; however, risk measures are most often described in terms of the mean value of a probability distribution function (pdf) wherein the uncertainty on the pdf is characterized by the 5th and 95th percentile values. As such, this discussion assumes that risk metrics are described in terms of the mean value of the risk measure pdf. The comparison of the risk metric to the acceptance guideline is expressed as the numerical difference between the mean value of the risk metric and the acceptance guideline.

Although the 5th and 95th percentiles of the risk measure pdf are useful for understanding the impact of parameter uncertainty, the mean value of the risk metric is the primary value considered when the staff performs the parameter uncertainty review. For example, if the comparison of the risk metric and the acceptance guideline falls into Regime 2, as discussed in Section 9.1, but the 95th percentile falls into the Regime 3 regime, the staff would characterize the comparison as being in Regime 2.

Once the comparison of the risk metric to the acceptance guideline has been characterized, the staff then determines the significance of the comparison. Similar to the process of determining the justification needed for an application in Stage F, the relative significance of the comparison increases as the risk metric approaches, challenges, or exceeds the acceptance guidelines.

If the comparison of the mean value of risk measure and the acceptance guideline is not significant (e.g., the comparison is characterized as Regime 1), the staff continues on with the risk-informed review process. If the comparison is significant, the staff proceeds to Step G-3b to determine whether the acceptability of the proximity of the risk measure to the acceptance guidelines has been adequately justified.

Step G-3b: ***Determine if there is Adequate Justification for the Acceptability of the Proximity of the Risk metric to the Acceptance Guideline***

The next step in the parameter uncertainty review involves determining whether there is adequate justification for the acceptability of the proximity of the risk metric to the acceptance guideline. The staff accomplishes this by determining whether the licensee has demonstrated an adequate understanding of the significance of the proximity and whether adequate justification has been provided for its acceptability. The staff's review includes but is not limited to a review of the application's technical bases, supporting analyses, justifications, and any other information provided by the licensee to support the acceptability of the application. In general, the staff uses a higher degree of scrutiny and requires more justification as the risk metric approaches, challenges, or exceeds the acceptance guideline.

If the staff determines that there not adequate justification for the proximity of the risk metric to the acceptance guideline, the application may be rejected as a risk-informed application, but not before the staff has performed a review of the licensee's assessment of model uncertainty. If the staff determines that there is adequate justification for the proximity of the risk metric to the acceptance guideline, the staff continues on with the risk-informed review process.

9.4   Risk-Informed Review of Model Uncertainty

The purpose of this section is to describe the process the staff uses to determine whether a licensee's treatment of model uncertainty associated with the PRA is acceptable for a given risk-informed application. This determination is made by assessing whether the licensee has adequately identified sources of model uncertainty and related assumptions that are key to the decision and, if so, whether adequate justification is provided for the acceptability of the application given the impact of those key sources of modeling uncertainty and related assumptions.

As discussed in Section 7, the licensee's process of identifying key sources of model uncertainty and related assumptions is dependent on whether the application risk metrics challenge or exceed the acceptance guidelines. In particular, when the parameter uncertainty assessment does not result in the risk metrics challenging or exceeding the acceptance guidelines, the sources of model uncertainty and related assumptions that are relevant to the application may be further screened to identify only those that are key to the decision. When there is a challenge to or exceedance of the acceptance guidelines, all of the sources of model uncertainty and related assumptions that are relevant to the application are treated as if they were key to the application. Figure 9-5 illustrates the steps of the model uncertainty review process.

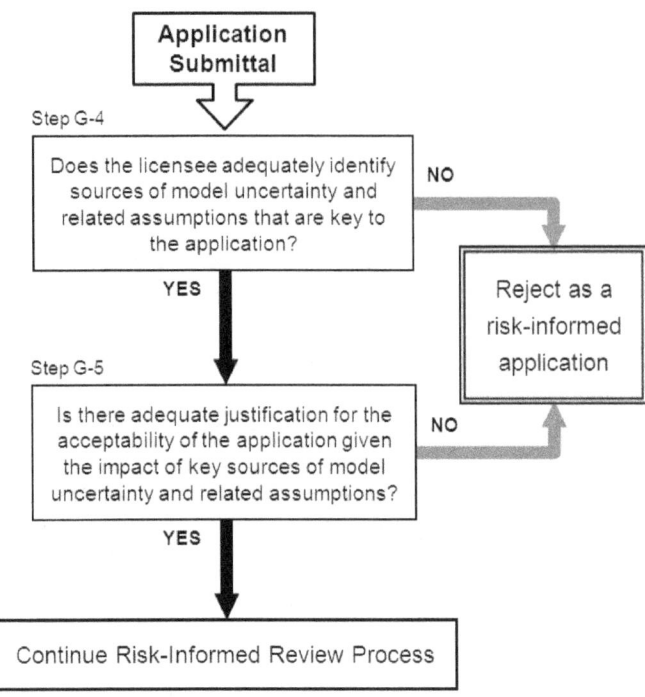

Figure 9-4 Overview of the risk-informed review of model uncertainty

As shown above, the staff first determines whether the licensee has adequately identified the sources of model uncertainty and related assumptions that are key to the application (i.e., those that may influence the decision being made). Next, the staff determines whether adequate justification has been provided for the acceptability of the application given the impact of the identified key sources of model uncertainty and related assumptions. These two steps are discussed below in detail.

Step G-4: _Adequate Identification of Key Sources of Model Uncertainty and Related Assumptions_

The first step performed by the staff in the model uncertainty review involves determining whether the licensee has adequately identified the sources of model uncertainty and related assumptions that are key to the decision. This step starts with a review of all the sources of model uncertainty and related assumptions identified by the licensee to determine whether the licensee has identified all sources of model uncertainty and related assumptions from the base PRA from which the application PRA has been derived.

Following this review, the staff determines whether the licensee has appropriately identified the sources of model uncertainty and related assumptions that are relevant to the decision. In making this determination, the staff assesses whether the licensee (1) demonstrates an adequate understanding of the way in which the PRA is used to support the application; (2) adequately identifies sources of model uncertainty and related assumptions in the base PRA that are relevant to the PRA results needed for the application.

Next, the staff determines whether the identified relevant sources of model uncertainty and related assumptions have been appropriately characterized. For each relevant source of model uncertainty or related assumption, this determination involves assessing whether the licensee demonstrates an adequate understanding of (1) the part of the PRA that is affected; (2) the modeling approach or assumptions used; (3) the impact on the PRA; and (4) whether there is an associated conservatism bias.

After the staff determines that the relevant sources of model uncertainty and related assumptions have been appropriately characterized, the staff then determines whether the licensee has appropriately identified those sources of model uncertainty and related assumptions that are key to the decision. As discussed in Section 7, the licensee's process of identifying key sources of model uncertainty and related assumptions is dictated by whether the application risk metrics have already been shown to meet the acceptance guidelines (i.e., the acceptance guidelines are not challenged or exceeded). As such, the staff first determines whether the licensee's parameter uncertainty assessment demonstrates that the application risk metrics challenge or exceed the acceptance guidelines.

When the parameter uncertainty assessment demonstrates that the risk metrics do not challenge or exceed the acceptance guidelines, the licensee then quantitatively assesses each of the relevant sources of model uncertainty and related assumptions to determine which ones are key to the decision. Based on the definition of a key source of model uncertainty or related assumption, assessing the quantitative impact of the relevant sources of model uncertainty and related assumptions involves performing sensitivity analyses to determine whether the application acceptance guidelines are, at a minimum, challenged. As such, the staff reviews the licensee's sensitivity analysis for each relevant source of model uncertainty or related assumption to determine the following:

- the sensitivity analysis has been performed in a technically correct manor

- the sensitivity analysis adequately accounts for the impact of and demonstrates whether the source of model uncertainty or related assumption could, at a minimum, potentially result in a challenge to the acceptance guidelines

- if there is a potential challenge to the acceptance guidelines, the licensee has appropriately identified the source of model uncertainty or related assumption as being key to the decision

When the parameter uncertainty assessment demonstrates that the risk metrics do challenge or exceed the acceptance guidelines, each relevant source of model uncertainty or related assumption is treated as being key to the decision. As such, the staff reviews the licensee's sensitivity analysis for each relevant source of model uncertainty or related assumption to determine whether the analysis:

- has been performed in a technically correct manor
- adequately accounts for the impact on the risk metrics

If the staff determines that the licensee adequately identified the key sources of model uncertainty and related assumptions, the staff review proceeds to Step G-5 wherein the staff determines whether the acceptability of the application is adequately justified given the impact of the identified key sources of model uncertainty and related assumptions. If the staff determines that the licensee has not adequately identified the key sources of model uncertainty, the application may be rejected as a risk-informed application.

Step G-5: *Adequacy of the Acceptability of the Risk-Informed Application Results*

In this Step, the staff determines whether adequate justification has been provided for the acceptability of the risk-informed application results, given the impact of the key sources of model uncertainty and related assumptions. The staff accomplishes this by reviewing the licensee's argument for the acceptability application. This may include, but is not limited to, a review of the application technical bases supporting analyses, compensatory measures or monitoring requirements, and other qualitative considerations.

If it is determined that adequate justification has been provided for the acceptability of the risk-informed application results, the staff continues on with the risk-informed review process. If not, the application may be rejected as a risk-informed application.

9.5   Risk-informed Review Process

The review process described in Sections 9.2, 9.3, and 9.4 addresses the first three of the following questions:

- How do the risk results compare to the acceptance guidelines?
- Is the scope and level of detail of the PRA appropriate for the application?
- Is the PRA model technically adequate?
- Is the acceptability of the application adequately justified?

Although these questions have been addressed sequentially, the staff review of a risk-informed application actually begins with the comparison of the application risk results to the acceptance guidelines. The justification needed to demonstrate the acceptability of a given risk-informed application is largely dictated by the proximity of the risk results to the acceptance guidelines. In general, an application can be characterized as falling into one of the following four general regimes based on the proximity of the risk results to the acceptance guidelines:

- Regime 1—The risk results are well below the acceptance guidelines
- Regime 2—The risk results are closer to, but do not challenge the acceptance guidelines
- Regime 3—The risk results challenge the acceptance guidelines
- Regime 4—The risk results clearly exceed the acceptance guidelines

How the first three questions above are addressed in the context of the acceptance guidelines is described below.

Regime 1: The risk results are well below the acceptance guidelines

An application may be characterized as being in Regime 1 when the application PRA risk metrics are well below the acceptance guidelines. The risk metrics are considered to be well below the acceptance guidelines when the mean value of the risk metrics are less than the acceptance guidelines by approximately one order of magnitude or more, as illustrated in Figure 9-1 by the white area.

For applications that fall into Regime 1, the staff would look for either a qualitative or quantitative assessment which demonstrates that the state-of-knowledge correlation (SOKC) does not impact the results of the PRA. It is important to consider that the SOKC may have some impact on application-specific scenarios that use ΔCDF as a risk metric. In these cases, the staff would evaluate whether the application considers the impact of the SOKC on ΔCDF. The staff would also evaluate the application to determine whether the validity of the assumptions made in the application PRA will be appropriately monitored via the implementation of specific measures and criteria. Specifically, the staff would be looking for whether the measures are appropriate for the application and that the criteria are set at reasonable limits. Moreover, the staff would look to see whether degraded performance can be detected in a timely fashion. The staff would likely place little importance on the licensee's use of compensatory measures, depending on the justification that is provided. The staff would review the peer review findings of the licensee's PRA to identify any findings of particular relevance to the application. Finally, the staff would generally not perform an audit on the application PRA when an application is in Regime 1.

Regime 2: The risk results are close to, but do not challenge the acceptance guidelines

An application may be characterized as being in Regime 2 when the application PRA risk metrics are closer to the acceptance guidelines than they are in Regime 1, but do not challenge the application acceptance guidelines. The risk metrics are considered to be close to the acceptance guidelines when the mean value of the risk metrics are approximately within an order of magnitude of, but do not approach the acceptance guidelines, as illustrated in Figure 9-1 by the light gray area.

For applications that fall into Regime 2, the staff would look for an assessment which shows that the SOKC does not impact the results of the PRA. As described for Regime 1, the staff would examine the application to ensure that the proposed performance monitoring is appropriate and adequate for the application and whether degraded performance can be detected in a timely fashion.

The staff would examine the peer review findings with a higher degree of scrutiny than for applications that fall into Regime 1 so as to better understand how particular findings were resolved as well as the general impact of the findings. In general, it is unlikely the staff would perform an audit on the application PRA for those applications that fall into Regime 2.

Regime 3: The risk results challenge the acceptance guidelines

An application may be characterized as being in Regime 3 when the application PRA risk metrics challenge the acceptance guidelines. The risk metrics are considered to challenge the acceptance guidelines when the mean value of the risk metrics are either slightly less than or slightly greater than the acceptance guidelines, as illustrated in Figure 9-1 by the dark gray area.

For applications that fall into Regime 3, the staff expects that a quantitative assessment would be provided which shows that the SOKC does not impact the results of the PRA. As described for Regime 1, the staff would examine the application to ensure that the proposed performance monitoring is adequate and determine whether degraded performance can be detected in a timely fashion. Applications that fall in this regime are expected to have compensatory measures in place. The staff would examine the peer review findings with an even higher degree of scrutiny than for applications that fall into Regime 2 so as to better understand how particular findings were resolved as well as the general impact of the findings. Additionally, the staff would perform an audit of the application PRA to determine the cause of the change in risk and would generally only be limited to an investigation of the significant issues in the PRA.

Regime 4: The risk results exceed the acceptance guidelines

An application may be characterized as being in Regime 4 when the application PRA risk metrics significantly exceed the acceptance guidelines. The risk measures estimates are considered to significantly exceed the acceptance guidelines when the mean value of the risk metrics are more than just slightly greater than the acceptance guidelines, as illustrated in Figure 9-1 by the black area.

For applications that fall into Regime 4, the staff would expect the licensee to clearly demonstrate that the PRA results are conservative and bounding. In making this determination, the staff would (1) look for a quantitative assessment which shows that the SOKC does not impact the results of the PRA; (2) examine the application to ensure that the proposed performance monitoring is adequate and determine whether degraded performance can be detected in a timely fashion; (3) determine the appropriateness of the compensatory measures; (4) thoroughly review the licensee's PRA peer review findings; and (5) perform a more in-depth audit of the application PRA than would be performed if the application were in Regime 3.

In general, the above characterization scheme relates to typical risk-informed applications and may not necessarily apply to more unique situations. For example, an application that potentially significantly exceeds the acceptance guidelines and, therefore would fall into Regime 4, may also represent an overall reduction in plant risk. In this example, the above characterization scheme may not be applicable.

In many cases, a risk-informed application requires consideration of the risk impact from multiple hazard groups and plant operational states and, as such, the risk contribution from each analysis must be combined, or aggregated, into a single risk measure, such as core damage frequency or large early release frequency. As such, the staff determines whether the licensee demonstrates adequate understanding of both the level of detail associated with the modeling of each of the hazard groups as well as the hazard group-specific parameter and model uncertainties. This involves determining whether the licensee demonstrates an adequate understanding of (1) the individual risk contributions and the parameter uncertainty associated with the analysis of each hazard group or plant operational state and (2) the sources and effects of conservatisms and model uncertainties that significantly impact the application results.

If accepted by the staff, the risk-informed application is considered to have (1) an acceptable treatment of uncertainties and (2) met the fourth risk-informed decisionmaking principle of posing an acceptable risk impact to the plant (see Section 2.3). Conversely, if the staff rejects the application, the risk-informed application is considered to have an unacceptable treatment of uncertainties or poses an unacceptable risk impact to the plant.

The following is a brief summary description of the staff's review of the different types of uncertainty associated with PRAs in risk-informed decisionmaking.

- Completeness Uncertainty Review: The staff's completeness uncertainty review addresses (1) the adequacy of the scope and level of detail and (2) the technical acceptability of the licensee's PRA model including any screening analyses used to address missing PRA scope items. The staff addresses the adequacy of the scope and level of detail by comparing the scope and level of detail required by the specific risk-informed activity with the scope and level of detail of the submitted application PRA. This comparison identifies those PRA scope and level-of-detail items that have either not been modeled in the PRA or have been intentionally excluded for that application. The staff reviews the screening analyses provided by the licensee to determine if there is adequate justification for the exclusion of PRA items that are missing from the application. The staff then assesses the technical adequacy of the PRA used in the risk-informed application to determine whether any technical inadequacies are significant.

- Parameter Uncertainty Review: The staff's parameter uncertainty review addresses (1) whether the licensee has demonstrated an adequate understanding of the impact of parameter uncertainties (i.e., the significance of state-of-knowledge correlation) and (2) whether the licensee has provided adequate justification for the comparison between the risk metrics and the acceptance guidelines. Since both parameter and model uncertainties may impact the application risk results, the staff's determination of the acceptability of the risk metrics as compared to the acceptance guidelines necessarily includes a review of model uncertainty.

- Model Uncertainty Review: The staff's model uncertainty review addresses (1) whether the licensee has adequately identified key sources of model uncertainty and related assumptions; (2) whether the licensee has demonstrated an adequate understanding of the impact on the risk metrics due to key model uncertainties or related assumptions; and (3) whether the licensee has provided adequate justification for the acceptability of the comparison between the risk metrics and the acceptance guidelines. The staff reviews the application's technical bases and sensitivity analyses to determine whether the licensee has provided adequate justification for the acceptability of the impact of key sources of model uncertainty and related assumptions.

9.6   Summary of Stage G

The staff's risk-informed review is the last stage in the process of assessing the treatment of uncertainty associated with a licensee's PRA in a risk-informed application. The purpose of this section is to provide a description of the risk-informed review process, as it relates to the licensee's treatment of uncertainties. This process is used by the staff to determine whether a licensee has demonstrated an acceptable treatment of uncertainties associated with the PRA.

10. REFERENCES

[Apostolakis, 1981]   Apostolakis, G., and Kaplan, S., "Pitfalls in Risk Calculations," *Reliability Engineering,* Vol. 2, pp. 135-145, 1981.

[Apostolakis, 1994]   Apostolakis, G., "A Commentary on Model Uncertainty", in Proceedings of Workshop 1 in Advanced Topics in Risk and Reliability Analysis, Model Uncertainty, its Characterization and Quantification, NUREG/CP-0138, October, 1994.

[ACRS, 2003a]   Advisory Committee on Reactor Safeguards, Letter from M. Bonaca, ACRS Chairman, to Chairman Diaz, NRC, "Proposed Resolution of Public Comments on Draft Regulatory Guide (DG)-1122, 'An Approach for Determining the Technical Adequacy of Probabilistic Risk Results for Risk-Informed Decision Making," Washington, D.C., April 21, 2003 (Available in ADAMS via accession number ML031130260).

[ACRS, 2003b]   Advisory Committee on Reactor Safeguards, Letter from M. Bonaca, ACRS Chairman, to Chairman Diaz, NRC, "Improvement of the Quality of Risk Information for Regulatory Decision Making," Washington, D.C., May 16, 2003 (Available in ADAMS via accession number ML031420832).

[ASME/ANS, 2009]   ASME/American Nuclear Society, "Standard for Level 1/Large Early Release Frequency Probabilistic Risk Assessment for Nuclear Power Plant Applications," ASME/ANS RA-Sa-2009, March 2009.

[EPRI, 1985]   Electric Power Research Institute, "Classification and Analysis of Reactor Operating Experience Involving Dependent Events," EPRI NP-3967, Palo Alto, CA, June 1985.

[EPRI, 1988]   Electric Power Research Institute, "Procedures for Treating Common Cause Failures in Safety and Reliability Studies," NUREG/CR-4780, EPRI NP-5613, PLG-0547, Vol. 1, Palo Alto, CA, January 1988.

[EPRI, 1995]   Electric Power Research Institute, "Probabilistic Safety Assessments Applications Guide," EPRI TR 105396, Palo Alto, CA, August 1995.

[EPRI, 2003]   Electric Power Research Institute, "Parametric Uncertainty Impacts on Option 2 Safety Significance Categorization," EPRI TR 1008905, Palo Alto, CA, June 2003.

[EPRI, 2004]   Electric Power Research Institute, "Guideline for the Treatment of Uncertainty in Risk-Informed Applications: Technical Basis Document," EPRI TR 1009652, Palo Alto, CA, December 2004.

[EPRI, 2005]   Electric Power Research Institute, "EPRI/NRC-RES Fire PRA Methodology for Nuclear Power Facilities," NUREG/CR-6850.

[EPRI, 2006]   Electric Power Research Institute, "Guideline for the Treatment of Uncertainty in Risk-Informed Applications: Applications Guide," EPRI 1013491, Palo Alto, CA, 2006.

[EPRI, 2008]   Electric Power Research Institute, "Treatment of Parameter and Model Uncertainty for Probabilistic Risk Assessments," EPRI 1016737, Palo Alto, CA, December 2008.

[EPRI, 2012]   Electric Power Research Institute, "Practical Guidance on the Use of PRA in Risk-Informed Applications with a Focus on the Treatment of Uncertainty," EPRI 1026511, Palo Alto, CA, 2012.

[Helton, 1996]   Helton, J.C., and Burmeister, D.E., (editors), "Treatment of Aleatory and Epistemic Uncertainty," special issue of *Reliability Engineering and System Safety,* Vol. 54, 2-3 (1996)

[INL, 1998]   Idaho National Engineering and Environmental Laboratory, "Common-Cause Failure Database and Analysis System: Event Data Collection, Classification, and Coding," NUREG/CR- 6268, Rev. 1, INEEL/EXT-97-00696, Idaho Falls, ID, June 1998 (Available in ADAMS via accession number ML072970404).

[LLNL, 1997]   Senior Seismic Hazard Analysis Committee (SSHAC), "Recommendations for Probabilistic Seismic Hazard Analysis: Guidance on Uncertainty and Use of Experts," UCRL-ID-122160, NUREG/CR-6372, Lawrence Livermore National Laboratory, April 1997 (Available in ADAMS via accession number ML080090003).

[Modarres, 2006] M. Modarres, "Risk Analysis in Engineering: Techniques, tools, and Trends," CRC Press, Taylor & Francis Group, Boca Raton, Florida, 2006.

[NEI, 2005a]   Nuclear Energy Institute, "Process for Performing Follow-on PRA Peer Reviews Using the ASME PRA Standard," NEI-05-04, Washington, D.C., January 2005 (Available in ADAMS via accession number ML083430462).

[NEI, 2005b]   Nuclear Energy Institute, "10 CFR 50.69 SSC Categorization Guideline", NEI-00-04, Washington, D.C., July 2005. (Available in ADAMS via accession number ML052910035).

[NEI, 2006]   Nuclear Energy Institute, "Probabilistic Risk Assessment (PRA) Peer Review Process Guidance," Update to Revision 1, NEI-00-02, Washington, D.C., October 2006 (Available in ADAMS via accession number ML061510619).

[NRC, 1975]   U.S. Nuclear Regulatory Commission, "Standard Review Plan for the Review of Safety Analysis Reports for Nuclear Power Plants," NUREG-0800, Washington, D.C., November 1975.

[NRC, 1986]   U.S. Nuclear Regulatory Commission, "Safety Goals for the Operations of Nuclear Power Plants: Policy Statement," *Federal Register*, Vol. 51, p. 30028 (51 FR 30028), Washington, D.C., August 4, 1986.

[NRC, 1990]   U.S. Nuclear Regulatory Commission, "Severe Accident Risk: An Assessment for Five U.S. Nuclear Power Plants," NUREG-1150, Washington, DC, December 1990 (Available in ADAMS via accession number ML120960691).

[NRC, 1991]   U.S. Nuclear Regulatory Commission, NUREG-1407, "Procedural and Submittal Guidance for the Individual Plant Examination of External Events (IPEEE) for Severe Accident Vulnerabilities," Washington, D.C., June 1991 (Available in ADAMS via accession number ML063550238).

[NRC, 1995]   U.S. Nuclear Regulatory Commission, "Final Policy Statement 'Use of Probabilistic Risk Assessment (PRA) Methods in Nuclear Regulatory Activities," *Federal Register*, Vol. 51, p. 42622 (51 FR 42622), Washington, D.C., 1995.

[NRC, 1996]   U.S. Nuclear Regulatory Commission, "Branch Technical Position on the Use of Expert Elicitation in the High-Level Radioactive Waste Program,"NUREG-1563, Washington, DC, November 1996 (Available in ADAMS via accession number ML033500190).

[NRC, 1997a]   U.S. Nuclear Regulatory Commission, "Acceptance Guidelines and Consensus Standards for Use in Risk-Informed Regulation," SECY-97-221, Washington, D.C., September 30, 1997.

[NRC, 1997b]   U.S. Nuclear Regulatory Commission, "Final Regulatory Guidance on Risk-Informed Regulation: Policy Issues," SECY-97-287, Washington, D.C., December 12, 1997.

[NRC, 1998a]   U.S. Nuclear Regulatory Commission, "An Approach for Plant-Specific, Risk-Informed Decision Making: Inservice Testing," RG 1.175, Washington, D.C., August 1998 (Available in ADAMS via accession number ML003740149).

[NRC, 1998b]   U.S. Nuclear Regulatory Commission, "An Approach for Plant-Specific, Risk-Informed Decision Making: Technical Specifications," RG 1.177, Rev. 1, Washington, D.C., May 2011 (Available in ADAMS via accession number ML100910008).

[NRC, 1999]   U.S. Nuclear Regulatory Commission, "Staff Requirements Memorandum - SECY-98-144 - White Paper on Risk-Informed and Performance-Based Regulation," SRM-SECY-98-144, Washington, D.C., March 1, 1999.

[NRC, 2002]   U.S. Nuclear Regulatory Commission, "An Approach for Using Probabilistic Risk Assessment in Risk-Informed Decisions on Plant-Specific Changes to the Licensing Basis," RG 1.174, Rev. 2, Washington, D.C., May 2011 (Available in ADAMS via accession number ML100910006).

[NRC, 2003a]   U.S. Nuclear Regulatory Commission, Letter from William Travers, Executive Director for Operations, to M. Bonaca, ACRS Chairman, "Proposed Resolution of Public Comments on Draft Regulatory Guide

(DG)-1122, 'An Approach for Determining the Technical Adequacy of Probabilistic Risk Results for Risk-Informed Decision Making'," Washington, D.C., June 4, 2003 (Available in ADAMS via accession number ML031410189).

[NRC, 2003b]   U.S. Nuclear Regulatory Commission, Letter from William Travers, Executive Director for Operations, to M. Bonaca, ACRS Chairman, "Improvement of the Quality of Risk Information for Regulatory Decision Making," Washington, D.C., August 4, 2003 (Available in ADAMS via accession number ML031980447).

[NRC, 2003c]   U.S. Nuclear Regulatory Commission, "An Approach for Plant-Specific Risk-Informed Decision Making for Inservice Inspection of Piping," RG 1.178, Washington, D.C., September 2003 (Available in ADAMS via accession number ML032510128).

[NRC, 2003d]   U.S. Nuclear Regulatory Commission, "Safety Evaluation of Topical Report WCAP-15603, Revision 1, 'WOG 2000 Reactor Coolant Pump Seal Leakage Model for Westinghouse PWRs," Washington, D.C., May 30, 2003 (Available in ADAMS via accession number ML031400376).

[NRC, 2003e]   U.S. Nuclear Regulatory Commission, "PRA Quality Expectations and Requirements," Staff Requirements Memorandum COMNJD-03-0002, Washington D.C., December 18, 2003 (Available in ADAMS via accession number ML033520457).

[NRC, 2004]   U.S. Nuclear Regulatory Commission, "Regulatory Analysis Guidelines of the U.S. Nuclear Regulatory Commission (rev. 4)", NUREG/BR-0058, Washington, D.C., September 2004 (Available in ADAMS via accession number ML042820192).

[NRC, 2005]   U.S. Nuclear Regulatory Commission, "Good Practices for Implementing Human Reliability Analysis," NUREG-1792, Washington, D.C., April 2005 (Available in ADAMS via accession number ML050950060).

[NRC, 2006a]   U.S. Nuclear Regulatory Commission, "Guidelines for Categorizing Structures, Systems, and Components in Nuclear Power Plants According to Their Safety Significance," RG 1.201, Washington, D.C., May 2006 (Available in ADAMS via accession number ML061090627).

[NRC, 2006b]   U.S. Nuclear Regulatory Commission, "Risk-Informed, Performance-Based Fire Protection for Existing Light-Water Nuclear Power Plants," RG 1.205, Washington, D.C., May 2006 (Available in ADAMS via accession number ML092730314).

[NRC, 2007a]   U.S. Nuclear Regulatory Commission, "An Approach for Determining the Technical Adequacy of Probabilistic Risk Assessment Results for Risk-Informed Activities," RG 1.200, Revision 2, Washington, D.C., March 2009 (Available in ADAMS via accession number ML090410014).

[NRC, 2007b]   U.S. Nuclear Regulatory Commission, "Combined License Applications for Nuclear Power Plants (LWR Edition)" RG 1.206, Washington, D.C., June 2007 (Available in ADAMS via accession number ML070720184).

[NRC, 2007c]   U.S. Nuclear Regulatory Commission, "Standard Review Plan, Section 19.0, 'Probabilistic Risk Assessment and Severe Accident Evaluation for New Reactors,'" Revision 2, Washington, D.C., June 2007 (Available in ADAMS via accession number ML071700652).

[Reinert, 2006] Reinert, J. M., and Apostolakis, G. E., "Including Model Uncertainty in Risk-informed Decision Making," Annals of Nuclear Energy, Vol. 33, pp. 354-369, January 19, 2006.

[SNL, 1992a]   Sandia National Laboratories, "Analysis of the LaSalle Unit 2 Nuclear Power Plant: Risk Methods Integration and Evaluation Program (RMIEP), Internal Fire Analysis" SAND92-0537, NUREG/CR-4832 Vol. 9, Albuquerque, NM, October 1992 (Available in ADAMS via accession number ML073470007).

[SNL, 1992b]   Sandia National Laboratories, "Methods for External Event Screening Quantification: Risk Methods Integration Evaluation Program (RMIEP) Methods Development," SAND87-7156, NUREG/CR-4839, Albuquerque, NM, July 1992 (Available in ADAMS via accession number ML062260210).

[SNL, 2003]   Sandia National Laboratories, "Handbook of Parameter Estimation for Probabilistic Risk Assessment," SAND2003-3348P, NUREG/CR-6823, Albuquerque, NM, September 2003 (Available in ADAMS via accession number ML032900131).

[Zio, 1996]   E. Zio and G. Apostolakis, "Two Methods for the Structured Assessment of Model Uncertainty by Experts in Performance Assessments of Radioactive Waste Repositories," Reliability Engineering and System Safety, Vol. 54, no. 2-3, pp. 225–241, December 1996.

NRC FORM 335 (12-2010) NRCMD 3.7	U.S. NUCLEAR REGULATORY COMMISSION	1. REPORT NUMBER (Assigned by NRC, Add Vol., Supp., Rev., and Addendum Numbers, if any.)
	BIBLIOGRAPHIC DATA SHEET *(See instructions on the reverse)*	NUREG-1855, Rev. 1

2. TITLE AND SUBTITLE	3. DATE REPORT PUBLISHED	
Guidance on the Treatment of Uncertainties Associated with PRAs in Risk-Informed Decisionmaking	MONTH	YEAR
	March	2013
Draft for Comment	4. FIN OR GRANT NUMBER	

5. AUTHOR(S)	6. TYPE OF REPORT
M. Drouin, A. Gilbertson, Office of Nuclear Regulatory Research G. Parry, Former NRC Employee J. Lehner, G. Martinez-Guridi, Brookhaven National Laboratory J. LaChance, T. Wheeler, Sandia National Laboratories	NUREG, Draft
	7. PERIOD COVERED (Inclusive Dates)

8. PERFORMING ORGANIZATION - NAME AND ADDRESS (If NRC, provide Division, Office or Region, U. S. Nuclear Regulatory Commission, and mailing address; if contractor, provide name and mailing address.)

Division of Risk Analysis Brookhaven National Laboratory Sandia National Laboratories
Office of Nuclear Regulatory Research Upton, New York 11973 Albuquerque, New Mexico 87185
U. S. Nuclear Regulatory Commission
Washington, DC 20555-0001

9. SPONSORING ORGANIZATION - NAME AND ADDRESS (If NRC, type "Same as above", if contractor, provide NRC Division, Office or Region, U. S. Nuclear Regulatory Commission, and mailing address.)

Division of Risk Assessment
Office of Nuclear Regulatory Research
U. S. Nuclear Regulatory Commission
Washington, DC 20555-0001

10. SUPPLEMENTARY NOTES
This revision supersedes the older report

11. ABSTRACT (200 words or less)

This document provides guidance on how to treat uncertainties associated with probabilistic risk assessment (PRA) in risk-informed decisionmaking. The objectives of this guidance include fostering an understanding of the uncertainties associated with PRA and their impact on the results of PRA and providing a pragmatic approach to addressing these uncertainties in the context of the decisionmaking.

This document describes the characteristics of a risk model and, in particular, a PRA. Since the focus of this document is epistemic uncertainty (i.e., uncertainties in the formulation of the PRA model), it provides guidance on identifying and describing the different types of sources of epistemic uncertainty and the different ways that they are treated. The different types of epistemic uncertainty are parameter, model, and completeness uncertainties.

The final part of the guidance addresses the uncertainty in PRA results in the context of risk-informed decisionmaking and, in particular, the interpretation of the results of the uncertainty analysis when comparing PRA results with the acceptance guidelines established for a specified application. In addition, guidance is provided for addressing completeness uncertainty in risk-informed decisionmaking. Such consideration includes using a program of monitoring, feedback, and corrective action.

12. KEY WORDS/DESCRIPTORS (List words or phrases that will assist researchers in locating the report.)	13. AVAILABILITY STATEMENT
risk-informed, probabilistic risk assessment, uncertainty, parameter, model, sensitivity, bounding, decisionmaking, risk analysis, epistemic	unlimited
	14. SECURITY CLASSIFICATION
	(This Page) unclassified
	(This Report) unclassified
	15. NUMBER OF PAGES
	16. PRICE

NUREG-1855, Rev. 1
Draft Report

Guidance on the Treatment of Uncertainties Associated
with PRAs in Risk-Informed Decisionmaking

March 2013